Smart Meter Data: Privacy and Cybersecurity

Brandon J. Murrill
Legislative Attorney

Edward C. Liu
Legislative Attorney

Richard M. Thompson II
Legislative Attorney

February 3, 2012

Congressional Research Service

7-5700

www.crs.gov

R42338

CRS Report for Congress ————————

Prepared for Members and Committees of Congress

Summary

Fueled by stimulus funding in the American Recovery and Reinvestment Act of 2009 (ARRA), electric utilities have accelerated their deployment of smart meters to millions of homes across the United States with help from the Department of Energy's Smart Grid Investment Grant program. As the meters multiply, so do issues concerning the privacy and security of the data collected by the new technology. This Advanced Metering Infrastructure (AMI) promises to increase energy efficiency, bolster electric power grid reliability, and facilitate demand response, among other benefits. However, to fulfill these ends, smart meters must record near-real time data on consumer electricity usage and transmit the data to utilities over great distances via communications networks that serve the smart grid. Detailed electricity usage data offers a window into the lives of people inside of a home by revealing what individual appliances they are using, and the transmission of the data potentially subjects this information to interception or theft by unauthorized third parties or hackers.

Unforeseen consequences under federal law may result from the installation of smart meters and the communications technologies that accompany them. This report examines federal privacy and cybersecurity laws that may apply to consumer data collected by residential smart meters. It begins with an examination of the constitutional provisions in the Fourth Amendment that may apply to the data. As we progress into the 21st century, access to personal data, including information generated from smart meters, is a new frontier for police investigations. The Fourth Amendment generally requires police to have probable cause to search an area in which a person has a reasonable expectation of privacy. However, courts have used the third-party doctrine to deny protection to information a customer gives to a business as part of their commercial relationship. This rule is used by police to access bank records, telephone records, and traditional utility records. Nevertheless, there are several core differences between smart meters and the general third-party cases that may cause concerns about its application. These include concerns expressed by the courts and Congress about the ability of technology to potentially erode individuals' privacy.

If smart meter data and transmissions fall outside of the protection of the Fourth Amendment, they may still be protected from unauthorized disclosure or access under the Stored Communications Act (SCA), the Computer Fraud and Abuse Act (CFAA), and the Electronic Communications Privacy Act (ECPA). These statutes, however, would appear to permit law enforcement to access smart meter data for investigative purposes under procedures provided in the SCA, ECPA, and the Foreign Intelligence Surveillance Act (FISA), subject to certain conditions. Additionally, an electric utility's privacy and security practices with regard to consumer data may be subject to Section 5 of the Federal Trade Commission Act (FTC Act). The Federal Trade Commission (FTC) has recently focused its consumer protection enforcement on entities that violate their privacy policies or fail to protect data from unauthorized access. This authority could apply to electric utilities in possession of smart meter data, provided that the FTC has statutory jurisdiction over them. General federal privacy safeguards provided under the Federal Privacy Act of 1974 (FPA) protect smart meter data maintained by federal agencies, including data held by federally owned electric utilities.

A companion report from CRS focusing on policy issues associated with smart grid cybersecurity, CRS Report R41886, *The Smart Grid and Cybersecurity—Regulatory Policy and Issues*, by Richard J. Campbell, is also available.

Contents

Figures

Contacts

Overview

Smart meter technology is a key component of the Advanced Metering Infrastructure (AMI)[1] that will help the smart grid[2] link the "two-way flow of electricity with the two-way flow of information."[3] Privacy and security concerns surrounding smart meter technology arise from the meters' essential functions, which include (1) recording near-real time data on consumer electricity usage; (2) transmitting this data to the smart grid using a variety of communications technologies;[4] and (3) receiving communications from the smart grid, such as real-time energy prices or remote commands that can alter a consumer's electricity usage to facilitate demand response.[5]

Beneficial uses of AMI are developing rapidly, and like the early Internet, many applications remain unforeseen.[6] At a basic level, smart meters will permit utilities to "collect, measure, and analyze energy consumption data for grid management, outage notification, and billing purposes."[7] The meters may increase energy efficiency by giving consumers greater control over their use of electricity,[8] as well as permitting better integration of plug-in electric vehicles and renewable energy sources.[9] They may also aid in the development of a more reliable electricity grid that is better equipped to withstand cyber attacks and natural disasters, and help to decrease peak demand for electricity.[10] To be useful for these purposes, and many others, data recorded by

---

[1] AMI includes the meters at the consumer's residence or business, the communications networks that send data between the consumer and utility, and the data management systems that store and process data for the utility. ELECTRIC POWER RESEARCH INST., ADVANCED METERING INFRASTRUCTURE (AMI) (2007), *available at* http://www.ferc.gov/eventcalendar/Files/20070423091846-EPRI%20-%20Advanced%20Metering.pdf. The primary function of AMI is to "combine interval data measurement with continuously available remote communications" to increase energy efficiency and grid reliability, and decrease expenses borne by the utility and consumer. *Id.*

[2] The Energy Independence and Security Act of 2007 (EISA) lists ten characteristics of a smart grid. These include "[i]ncreased use of digital information and controls technology to improve reliability, security, and efficiency of the electric grid"; "[d]evelopment and incorporation of demand response, demand-side resources, and energy-efficiency resources"; and "[d]eployment of "smart" technologies (real-time, automated, interactive technologies that optimize the physical operation of appliances and consumer devices) for metering, communications concerning grid operations and status, and distribution automation." EISA, P.L. 110-140, §1301, 121 Stat. 1492, 1783-84 (2007) (to be codified at 42 U.S.C. §17381).

[3] DEP'T OF ENERGY, COMMUNICATIONS REQUIREMENTS OF SMART GRID TECHNOLOGIES 1 (2010) [hereinafter DEP'T OF ENERGY COMMUNICATIONS REPORT], *available at* http://energy.gov/sites/prod/files/gcprod/documents/ Smart_Grid_Communications_Requirements_Report_10-05-2010.pdf.

[4] *Id.* at 3, 5. These technologies include fiber optics, wireless networks, satellite, and broadband over power line. *Id.*

[5] *Id.* at 20. "Demand response is the reduction of the consumption of electric energy by customers in response to an increase in the price of electricity or heavy burdens on the system." *Id.*

[6] DEP'T OF ENERGY, DATA ACCESS AND PRIVACY ISSUES RELATED TO SMART GRID TECHNOLOGIES 5, 9 (2010) [hereinafter DEP'T OF ENERGY PRIVACY REPORT], *available at* http://energy.gov/sites/prod/files/gcprod/documents/ Broadband_Report_Data_Privacy_10_5.pdf; *see also* ELIAS LEAKE QUINN, SMART METERING & PRIVACY: EXISTING LAW AND COMPETING POLICIES: A REPORT FOR THE COLORADO PUBLIC UTILITIES COMMISSION 1, 12 (2009) [hereinafter COLORADO PRIVACY REPORT], *available at* http://www.dora.state.co.us/puc/docketsdecisions/DocketFilings/09I-593EG/09I-593EG_Spring2009Report-SmartGridPrivacy.pdf.

[7] DEP'T OF ENERGY COMMUNICATIONS REPORT, *supra* note 3, at 12.

[8] Companies are developing several new applications that use smart meter data to offer consumers and utilities better control over energy usage, for example by determining the energy efficiency of specific appliances within the household. DEP'T OF ENERGY PRIVACY REPORT, *supra* note 6, at 5, 9; *see also* COLORADO PRIVACY REPORT, *supra* note 6, at 1, 12.

[9] DEP'T OF ENERGY COMMUNICATIONS REPORT, *supra* note 3, at 1.

[10] *Id.* at 3.

---

smart meters must be highly detailed, and, consequently, it may show what individual appliances a consumer is using.[11] The data must also be transmitted to electric utilities—and possibly to third parties outside of the smart grid—subjecting it to potential interception or theft as it travels over communications networks and is stored in a variety of physical locations.[12]

These characteristics of smart meter data present privacy and security concerns that are likely to become more prevalent as government-backed initiatives expand deployment of the meters to millions of homes across the country. In the American Recovery and Reinvestment Act of 2009 (ARRA), Congress appropriated funds for the implementation of the Smart Grid Investment Grant (SGIG) program administered by the Department of Energy.[13] This program now permits the federal government to reimburse up to 50% of eligible smart grid investments, which include the cost to electric utilities of buying and installing smart meters.[14] In its annual report on smart meter deployment, the Federal Energy Regulatory Commission cited statistics showing that the SGIG program has helped fund the deployment of about 7.2 million meters as of September 2011.[15] At completion, the program will have partially funded the installation of 15.5 million meters.[16] By 2015, the Institute for Electric Efficiency expects that a total of 65 million smart meters will be in operation throughout the United States.[17]

Installation of smart meters and the communications technologies that accompany them may have unforeseen legal consequences for those who generate, seek, or use the data recorded by the meters. These consequences may arise under existing federal laws or constitutional provisions governing the privacy of electronic communications, data retention, computer misuse, foreign surveillance, and consumer protection. This report examines federal privacy and cybersecurity laws that may apply to consumer data collected by residential smart meters. It examines the legal implications of smart meter technology for consumers who generate the data, law enforcement officers who seek smart meter data from utilities, utilities that store the data, and hackers who access smart grid technology to steal consumer data or interfere with it. This report looks at federal laws that may pertain to the data when it is (1) stored in a utility-owned smart meter at a consumer's residence; (2) in transit between the meter and the smart grid by way of various communications technologies; and (3) stored on computers in the grid. This report does not address state or local laws, such as regulations by state Public Utilities Commissions, that may establish additional responsibilities for some electric utilities with regard to smart meter data. It also does not discuss the mandatory cybersecurity and reliability standards enforced by the North

---

[11] *See* NAT'L INST. OF STANDARDS AND TECH., GUIDELINES FOR SMART GRID CYBER SECURITY: VOL. 2, PRIVACY AND THE SMART GRID 14 (2010) [hereinafter NIST PRIVACY REPORT], *available at* http://csrc.nist.gov/publications/nistir/ir7628/nistir-7628_vol2.pdf.

[12] *Id.* at 3-4, 23-24, 29.

[13] The act provides $4.5 billion for "electricity delivery and energy reliability," which includes "activities to modernize the electric grid, to include demand responsive equipment," as well as "programs authorized under title XIII of the Energy Independence and Security Act of 2007." ARRA, P.L. 111-5, 123 Stat. 115, 138-39.

[14] ARRA §405(5), (8), 123 Stat. 115, 143-44 (amendment to be codified at 42 U.S.C. §17386) (amending the Energy Independence and Security Act of 2007 (EISA) to allow for the reimbursement of up to 50% of qualifying smart grid investments instead of only 20%); *see also* EISA, P.L. 110-140, §1306, 121 Stat. 1492, 1789-91 (to be codified as amended at 42 U.S.C. §17386) (initially establishing the SGIG program).

[15] FED. ENERGY REGULATORY COMM'N, ASSESSMENT OF DEMAND RESPONSE & ADVANCED METERING 3 (2011), *available at* http://www.ferc.gov/legal/staff-reports/11-07-11-demand-response.pdf.

[16] *Id.*

[17] INST. FOR ELECTRIC EFFICIENCY, UTILITY-SCALE SMART METER DEPLOYMENTS, PLANS & PROPOSALS 1 (2011), *available at* http://www.edisonfoundation net/iee/issuebriefs/SmartMeter_Rollouts_0911.pdf.

American Electric Reliability Corporation, which impose obligations on utilities that participate in the generation or transmission of electricity.[18]

General federal privacy safeguards provided under the Federal Privacy Act of 1974 (FPA) protect smart meter data maintained by federal agencies, including data held by federally owned electric utilities. Section 5 of the Federal Trade Commission Act (FTC Act) allows the Federal Trade Commission (FTC) to bring enforcement proceedings against electric utilities that violate their privacy policies or fail to protect meter data from unauthorized access, provided that the FTC has statutory jurisdiction over the utilities.

It is unclear how Fourth Amendment protection from unreasonable search and seizures would apply to smart meter data, due to the lack of cases on this issue. However, depending upon the manner in which smart meter services are presented to consumers, smart meter data may be protected from unauthorized disclosure or unauthorized access under the Stored Communications Act (SCA), the Computer Fraud and Abuse Act (CFAA), and the Electronic Communications Privacy Act (ECPA). If smart meter data is protected by these statutes, law enforcement would still appear to have the ability to access it for investigative purposes under procedures provided in the SCA, ECPA, and the Foreign Intelligence Surveillance Act (FISA).

Smart Meter Data: Privacy and Security Concerns

Residential smart meters present privacy and cybersecurity issues[19] that are likely to evolve with the technology.[20] In 2010, the National Institute of Standards and Technology (NIST) published a report identifying some of these issues, which fall into two main categories: (1) privacy concerns that smart meters will reveal the activities of people inside of a home by measuring their electricity usage frequently over time;[21] and (2) fears that inadequate cybersecurity measures surrounding the digital transmission of smart meter data will expose it to misuse by authorized and unauthorized users of the data.[22]

Detailed Information on Household Activities

Smart meters offer a significantly more detailed illustration of a consumer's energy usage than regular meters. Traditional meters display data on a consumer's *total* electricity usage and are typically read manually once per month.[23] In contrast, smart meters can provide *near real-time* usage data by measuring usage electronically at a much greater frequency, such as once every 15

---

[18] For additional information on the development of mandatory national smart grid privacy and cybersecurity standards by federal agencies, see MASS. INST. OF TECH., THE FUTURE OF THE ELECTRIC GRID 197-234 (2011) [hereinafter MIT GRID STUDY]; *see also* CRS Report R41886, *The Smart Grid and Cybersecurity—Regulatory Policy and Issues*, by Richard J. Campbell.

[19] According to the authors of the MIT study, cybersecurity "refers to all the approaches taken to protect data, systems, and networks from deliberate attack as well as accidental compromise, ranging from preparedness to recovery." MIT GRID STUDY, *supra* note 18, at 208. Closely related is the concept of "information privacy," which "deals with policy issues ranging from identification and collection to storage, access, and use of information." *Id.* at 219 n.viii.

[20] *See* NIST PRIVACY REPORT, *supra* note 11, at 1.

[21] *Id.* at 4, 11. Data that offers a high degree of detail is said to be "granular." *Id.*

[22] *See id.* at 4, 23-24, 29.

[23] *Id.* at 2, 9.

---

minutes.[24] Current smart meter technology allows utilities to measure usage as frequently as once every minute.[25] By examining smart meter data, it is possible to identify which appliances a consumer is using and at what times of the day, because each type of appliance generates a unique electric load "signature."[26] NIST wrote in 2010 that "research shows that analyzing 15-minute interval aggregate household energy consumption data can by itself pinpoint the use of most major home appliances."[27] A report for the Colorado Public Utilities Commission discussed an Italian study that used "artificial neural networks" to identify individual "heavy-load appliance uses" with 90% accuracy using 15-minute interval data from a smart meter.[28] Similarly, software-based algorithms would likely allow a person to extract the unique signatures of individual appliances from meter data that has been collected less frequently and is therefore less detailed.[29]

By combining appliance usage patterns, an observer could discern the behavior of occupants in a home over a period of time.[30] For example, the data could show whether a residence is occupied, how many people live in it, and whether it is "occupied by more people than usual."[31] According to the Department of Energy, smart meters may be able to reveal occupants' "daily schedules (including times when they are at or away from home or asleep), whether their homes are equipped with alarm systems, whether they own expensive electronic equipment such as plasma TVs, and whether they use certain types of medical equipment."[32] **Figure 1**, which appears in NIST's report on smart grid cybersecurity, shows how smart meter data could be used to decipher the activities of a home's occupants by matching data on their electricity usage with known appliance load signatures.

---

[24] *Id.* at 13.

[25] COLORADO PRIVACY REPORT, *supra* note 6, at 2. Some utilities may elect to receive data at less frequent intervals because "backhauling real-time or near real-time data from the billions of devices that may eventually be connected to the Smart Grid would require not only tremendous bandwidth" but also greater data storage capacities that could make the effort "economically infeasible." DEP'T OF ENERGY COMMUNICATIONS REPORT, *supra* note 3, at 20. However, the "trend" is for utilities to collect data more frequently. *See* COLORADO PRIVACY REPORT, *supra* note 6, at A-1 n.111.

[26] NIST PRIVACY REPORT, *supra* note 11, at 2, 14.

[27] *Id.* at 14. *But see* DEP'T OF ENERGY PRIVACY REPORT, *supra* note 6, at 9 (claiming, in 2010, that smart meter technology "cannot yet identify individual appliances and devices in the home in detail, but this will certainly be within the capabilities of subsequent generations of Smart Grid technologies").

[28] COLORADO PRIVACY REPORT, *supra* note 6, at 3 n.7, A-8.

[29] *Id.* at A-9.

[30] NIST PRIVACY REPORT, *supra* note 11, at 6 & n.9.

[31] *Id.* at 11.

[32] DEP'T OF ENERGY PRIVACY REPORT, *supra* note 6, at 2.

Figure 1. Identification of Household Activities from Electricity Usage Data
Unique Electric Load Signatures of Common Household Appliances

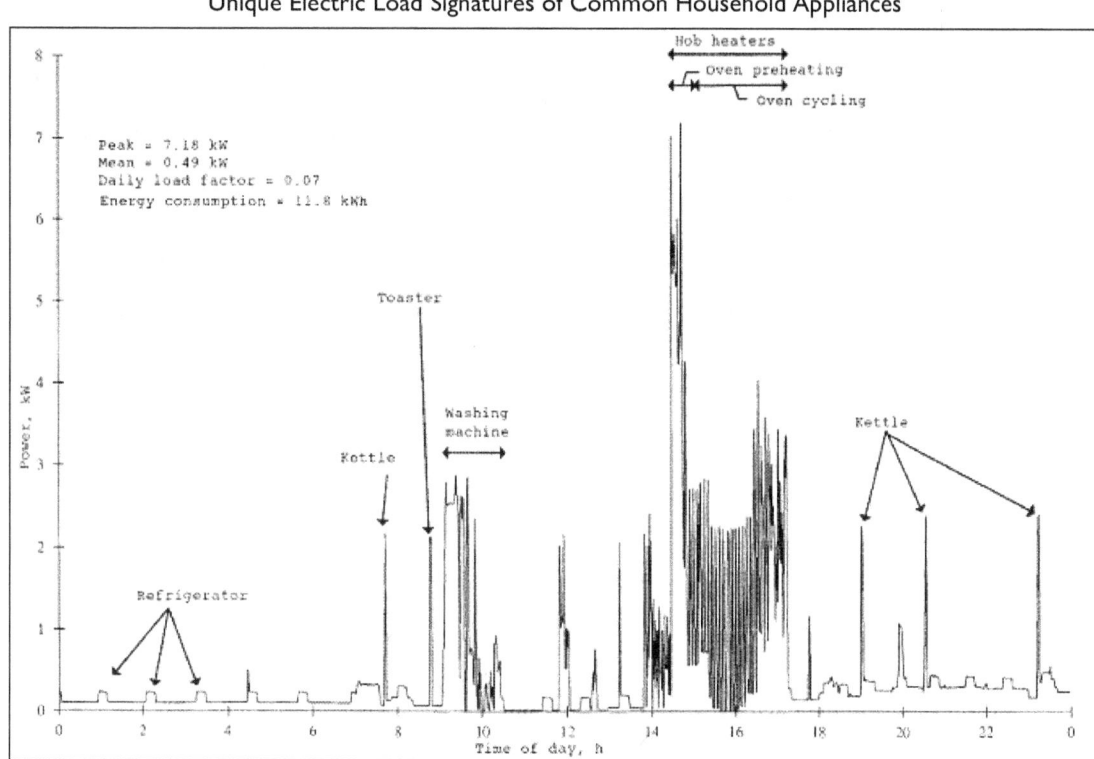

Source: NATIONAL INSTITUTE OF STANDARDS AND TECHNOLOGY (NIST), GUIDELINES FOR SMART GRID CYBER SECURITY: VOL. 2, PRIVACY AND THE SMART GRID 13 (2010), *available at* http://csrc.nist.gov/publications/nistir/ir7628/ nistir-7628_vol2.pdf.

Note: Researchers constructed this picture from electricity usage data collected at one-minute intervals using a nonintrusive appliance load monitoring (NALM) device, which is similar to a smart meter in the way that it records usage data. For a comparison of the technologies, see COLORADO PRIVACY REPORT, *supra* note 6, at A-1 to A-9.

Smart meter data that reveals which appliances a consumer is using has potential value for third parties, including the government. In the past, law enforcement agents have examined *monthly* electricity usage data from *traditional* meters in investigations of people they suspected of illegally growing marijuana.[33] For example, in *United States v. Kyllo*, a federal agent subpoenaed the suspect's electricity usage records from the utility and "compared the records to a spreadsheet for estimating average electrical use and concluded that Kyllo's electrical usage was abnormally high, indicating a possible indoor marijuana grow operation."[34] If law enforcement officers obtained near-real time data on a consumer's electricity usage from the utility company, their ability to monitor household activities would be amplified significantly.[35] For example, by observing when occupants use the most electricity, it may be possible to discern their daily schedules.[36]

---

[33] NIST PRIVACY REPORT, *supra* note 11, at 11, 29; *see also* United States v. Kyllo, 190 F.3d 1041, 1043 (9th Cir. 1999), *rev'd on other grounds*, 533 U.S. 27 (2001).

[34] *Kyllo*, 190 F.3d at 1043.

[35] *See supra* notes 26-32 and accompanying text.

[36] *See supra* note 32 and accompanying text.

As smart meter technology develops and usage data grows more detailed, it could also become more valuable to private third parties outside of the grid.[37] Data that reveals which appliances a person is using could permit health insurance companies to determine whether a household uses certain medical devices, and appliance manufacturers to establish whether a warranty has been violated.[38] Marketers could use it to make targeted advertisements.[39] Criminals could use it to time a burglary and figure out which appliances they would like to steal.[40] If a consumer owned a plug-in electric vehicle, data about where the vehicle has been charged could permit someone to identify a person's location and travel history.[41]

Even privacy safeguards, such as "anonymizing" data so that it does not reflect identity, are not foolproof.[42] By comparing anonymous data with information available in the public domain, it is sometimes possible to identify an individual—or, in the context of smart meter data, a particular household.[43] Moreover, a smart grid will collect more than just electricity usage data. It will also store data on the account holder's name, service address, billing information, networked appliances in the home, and meter IP address, among other information.[44] Many smart meters will also provide transactional records as they send data to the grid, which would show the time that the meter transmitted the data and the location or identity of the transmitter.[45]

Increased Potential for Theft or Breach of Data

Smart grid technology relies heavily on two-way communication to increase energy efficiency and reliability, including communication between smart meters and the utility (or other entity) that stores data for the grid.[46] Many different technologies will transmit data to the grid, including "traditional twisted-copper phone lines, cable lines, fiber optic cable, cellular, satellite, microwave, WiMAX, power line carrier, and broadband over power line."[47] Of these communications platforms, wireless technologies are likely to play a "prominent role" because they present fewer safety concerns and cost less to implement than wireline technologies.[48] According to the Department of Energy, a typical utility network has four "tiers" that collect and transmit data from the consumer to the utility.[49] These include "(1) the core backbone—the primary path to the utility data center; (2) backhaul distribution—the aggregation point for

---

[37] NIST PRIVACY REPORT, *supra* note 11, at 14, 35-36.

[38] *Id.* at 27-28.

[39] *Id.* at 28.

[40] *Id.* at 31.

[41] *Id.*

[42] *Id.* at 13.

[43] *See id.* at 13, 25.

[44] *Id.* at 26-27.

[45] *Id.* at 12 (drawing a comparison to telecommunications providers' "call detail records").

[46] *Id.* at 3; DEP'T OF ENERGY COMMUNICATIONS REPORT, *supra* note 3, at 3 (stating that "integrated two-way communications ... allows for dynamic monitoring of electricity use as well as the potential for automated electricity use scheduling."). As more consumers become generators of electricity through the use of "fuel cells, wind turbines, solar roofs, and the like," the importance of two-way communication will increase. MIT GRID STUDY, *supra* note 18, at 201.

[47] DEP'T OF ENERGY COMMUNICATIONS REPORT, *supra* note 3, at 3.

[48] *Id.* at 5, 51 n.215.

[49] *Id.* at 16.

neighborhood data; (3) the access point—typically the smart meter; and, (4) the HAN—the home network."[50] Energy usage data moves from the smart meter,[51] and then to an "aggregation point" outside of the residence such as "a substation, a utility pole-mounted device, or a communications tower."[52] The aggregation points gather data from multiple meters and "backhaul" it to the utility using fiber, T1, microwave, or wireless technology.[53] Utilities typically rely on their own private networks to communicate with smart meters because they have found these networks to be more reliable and less expensive than commercial networks.[54]

As NIST explains, consumer data moving through a smart grid becomes stored in many locations both within the grid and within the physical world.[55] Thus, because it is widely dispersed, it becomes more vulnerable to interception by unauthorized parties[56] and to accidental breach.[57] The movement of data also increases the potential for it to be stolen by unauthorized third parties while it is in transit, particularly when it travels over a wireless network[58]—or through communications components that may be incompatible with one another or possess outdated security protections.[59]

Smart Meters and the Fourth Amendment

The use of smart meters presents the recurring conflict between law enforcement's need to effectively investigate and combat crime and our desire for privacy while in our homes. With smart meters, police will have access to data that might be used to track residents' daily lives and routines while in their homes, including their eating, sleeping, and showering habits, what appliances they use and when, and whether they prefer the television to the treadmill, among a host of other details.[60] Though a potential boon to police, access to this data is not limitless. The Fourth Amendment, which establishes the constitutional parameters for government investigations, may restrict access to smart meter data or establish rules by which it can be obtained.[61] The Fourth Amendment ensures that the "right of the people to be secure in their persons, houses, papers, and effects, against unreasonable searches and seizures, shall not be violated...."[62] This section discusses whether the collection and use of smart meter data may

---

[50] *Id.*

[51] The home network will be used to provide *consumers* with near real-time data on their energy usage. *Id.* at 13-15.

[52] *Id.* Many urban installations use wireless mesh networks to carry data from the meters to the aggregation point. These networks are more reliable because each smart meter can serve as a router in the network, providing redundant network coverage. *Id.* at 18.

[53] *Id.* at 16, 19.

[54] *Id.* at 4, 19, 44.

[55] NIST PRIVACY REPORT, *supra* note 11, at 23.

[56] *Id.* at 23-24.

[57] *Id.* at 29.

[58] *See id.* at 9, 12, 33, and 36.

[59] MIT GRID STUDY, *supra* note 18, at 209, 213-16.

[60] Jack I. Lerner & Deirdre K. Mulligan, *Taking the "Long View" on the Fourth Amendment: Stored Records and the Sanctity of the Home*, 2008 STAN. TECH. L. REV. 3, ¶ 3 (2008).

[61] Additionally, as described below, there are federal statutory protections that may pertain to this data. State constitutional and statutory safeguards may also apply, but these are beyond the scope of this report.

[62] U.S. CONST. amend IV.

contravene this protection. Although there is no Fourth Amendment case on point, analogous cases may provide guidance.[63]

To assess whether there has been a Fourth Amendment violation, two primary questions must be asked: (1) whether there was state action; that is, was there sufficient government involvement in the alleged wrongdoing to trigger the Fourth Amendment; and (2) whether the person had an expectation of privacy that society is prepared to deem reasonable.[64] If the first question is answered in the affirmative, then the analysis moves to the second question. But if no state action is found, the analysis ends there and the Fourth Amendment does not apply. This subpart will first determine whether access to smart meter data by police, or by privately and publicly owned utilities, satisfies the state action doctrine, thereby warranting further Fourth Amendment review.

State Action: Privately Versus Publicly Owned Utilities

Most of the safeguards for civil liberties and individual rights contained in the U.S. Constitution apply only to actions by state and federal governments.[65] This rule, known as the state action doctrine, arises when a victim claims his constitutional rights have been violated, and therefore must prove the wrongdoer had sufficient connections with the government to warrant a remedy.[66] Applying the state action test is intended to determine whether a utility's collection and dissemination of smart meter data is governed by the Fourth Amendment, and if so, to what extent. Although there are many variations in the governance and ownership of utilities—some are privately owned, others publicly owned, some federally operated, and still others nonprofit cooperatives—they generally fall into two broad categories: public and private.[67] This section will analyze the constitutional differences between privately and publicly owned utilities under the state action doctrine and a public records theory.

Privately Owned and Operated Utilities

It is broadly said that the Fourth Amendment applies only to acts by the government.[68] But there are at least two exceptions to this rule. First, if a utility performs a function traditionally exercised by the government, it may be considered a state actor under the public function exception. Second, the Fourth Amendment may apply when a private utility acts as an instrument or agent of the police.[69]

---

[63] For additional analyses of smart meters under the Fourth Amendment, see Cheryl Dancey Balough, *Privacy Implications of Smart Meters*, 86 CHI.-KENT L. REV. 161 (2011); *see also* QUINN, *supra* note 6, at 28 ("[I]nterval data of electricity consumption appears to be in something of a no-man's-land under Supreme Court Fourth Amendment jurisprudence.").

[64] California v. Ciraolo, 476 U.S. 207, 211 (1986) (citing Katz v. United States, 389 U.S. 347, 360 (1967) (Harlan, J., concurring)).

[65] Civil Rights Cases, 109 U.S. 3, 11 (1883) ("It is State action of a particular character that is prohibited. Individual invasion of individual rights is not the subject-matter of the [Fourteenth] amendment."); *see* JOHN E. NOWAK & RONALD D. ROTUNDA, CONSTITUTIONAL LAW §12.1(a)(i) (8th ed. 2010).

[66] NOWAK & ROTUNDA, *supra* note 65.

[67] Determining whether a private actor is sufficiently "public" is not clear-cut. Then Justice Rehnquist noted, "[t]he true nature of the State's involvement may not be immediately obvious, and detailed inquiry may be required in order to determine whether the test is met." Jackson v. Metropolitan Edison Co., 419 U.S. 345, 351 (1974).

[68] Burdeau v. McDowell, 256 U.S. 465, 475 (1921).

[69] *See* United States v. Jacobsen, 466 U.S. 109, 113 (1984).

Under the public function exception, a nominally private entity is treated as a state actor when it assumes a role traditionally played by the government.[70] Determining when this exception applies has not proved easy,[71] but it is reasonably clear that private utilities do not, in most instances, satisfy it. In *Jackson v. Metropolitan Edison Co.*, a customer sued a privately owned utility under the Civil Rights Act of 1871 for improperly shutting off her service without providing her notice or a hearing.[72] The Supreme Court asked whether there was a close enough nexus between the state and the utility for the acts of the latter to be treated as those of the former.[73] Although the utility was heavily regulated by the state, it was held not to be a state actor.[74] The Court reasoned that the provision of utility service is not generally an "exclusive prerogative of the State."[75] Also absent was the symbiotic relationship between the utility and the state found in previous cases.[76] Though its holding was broad, the Court did not foreclose the possibility that a privately owned utility could be a state actor under different circumstances.[77] This possibility, however, appears narrow.

The Fourth Amendment may also apply to a private utility if its acts were directed by the government. Generally, searches performed by private actors without police participation or encouragement are not governed by the Fourth Amendment.[78] A search by a private insurance investigator, for instance, was not a "search" in the constitutional sense, though the evidence was ultimately used by the government at trial.[79] This result differs, however, if there is sufficient government involvement. If the search has been ordered or requested by the government, the private actor will become an "instrument or agent of the state" and must abide by Fourth Amendment strictures.[80] For example, the Fourth Amendment does not apply when a telephone company installs a pen register on its own initiative.[81] The same action constitutes a search, however, if requested by the government.[82]

This theory applies not only to direct instigation, but also on a broad, programmatic level. In the 1960s and 1970s the federal government required privately owned and operated airlines to institute new security measures to combat airline hijacking.[83] In *United States v. Davis*, the airline

---

[70] Marsh v. Alabama, 326 U.S. 501 (1946) (holding that privately owned property was equivalent to "community shopping center" thus private party was subject to the First and Fourteenth Amendments).

[71] *See* NOWAK & ROTUNDA, *supra* note 65, §12.2.

[72] *Jackson*, 419 U.S. at 347; *see also* Mays v. Buckeye Rural Elec. Coop., Inc., 277 F.3d 873, 880-81 (6th Cir. 2002) (holding that nonprofit cooperative utility was not a state actor under the federal constitution); Spickler v. Lee, No. 02-1954, 2003 U.S. App. LEXIS 6227, at *2 (1st Cir. March 31, 2003) (holding that private electric utility company was not a state actor).

[73] *Jackson*, 419 U.S. at 351.

[74] *Id.* at 358-59.

[75] *Id.* at 353.

[76] *Id.* at 357.

[77] *Id.* at 351.

[78] 1 WAYNE R. LAFAVE, SEARCH AND SEIZURE §1.8, at 255 (4th ed. 2004).

[79] United States v. Howard, 752 F.2d 220, 227-28 (6th Cir. 1985).

[80] Coolidge v. New Hampshire, 403 U.S. 443, 487 (1971) (internal quotation marks omitted); *see* LAFAVE, *supra* note 78, §1.8(b).

[81] United States v. Manning, 542 F.2d 685, 686 (6th Cir. 1976).

[82] People of Dearborn Heights v. Hayes, 82 Mich. App. 253, 258 (1978).

[83] United States v. Davis, 482 F.2d 893, 897-903 (9th Cir. 1973).

searched a passenger based on these requirements and found a loaded gun.[84] The Ninth Circuit held that it made no difference whether the search was conducted by a private or public official: "the search was part of the overall, nation-wide anti-hijacking effort, and constituted 'state action' for purposes of the Fourth Amendment."[85] Thus, if a private party is required to perform a search or collect data under federal or state laws or regulations, there will be sufficient state action for the Fourth Amendment to apply. Or, put another way, the government cannot circumvent the Fourth Amendment by requiring a private party to initiate a search or implement an investigative program.

This agency theory might apply to the collection of smart meter data. If the utility is accessing this information "independent of the government's intent to collect evidence for use in a criminal prosecution,"[86] the utility will not be considered an agent of the government for Fourth Amendment purposes. But there might be instances when government instigation will trigger further analysis. If, for example, the government requested the utility to record larger quantities of data than was customary (e.g., increasing the intervals from sub-15 minute intervals to sub-five minute or sub-one minute intervals), this would likely warrant Fourth Amendment scrutiny. Also, if the police requested the utility to hand over customer data, say, for spikes in energy commensurate with a marijuana growing operation, this would likely be a sufficient instigation to trigger further constitutional review. Other situations may arise where the government establishes a dragnet-type law enforcement scheme in which all smart meter data is filtered through police computers. This could also implicate the agency theory and warrant a finding of state action.

Publicly Owned and Operated Utilities

Although the Fourth Amendment (with its warrant and probable cause requirement) typically applies to public actors, in certain instances their collection of information may not fall under the Fourth Amendment or may prompt a lower evidentiary standard. The Supreme Court has infrequently considered the scope of the Fourth Amendment "on the conduct of government officials in noncriminal investigations,"[87] and even less frequently as to "noncriminal *noninvestigatory* governmental conduct."[88] Nonetheless, there are two lines of cases that may apply to smart meters in which the Fourth Amendment may not apply at all (noncriminal noninvestigatory conduct) or may be reduced (noncriminal investigations). The key to this analysis is the government's purpose in collecting the data.

The Supreme Court has developed a line of cases dubbed the "special needs" doctrine that permits the government to perform suspicionless searches if the special needs supporting the program outweigh the intrusion on the individual's privacy.[89] It is premised on the notion that "'special needs,' beyond the normal need for law enforcement, make the warrant and probable-cause requirement impracticable."[90] If, on the one hand, the objective of the search is not for law

---

[84] *Id.* at 895.

[85] *Id.* at 904.

[86] United States v. Howard, 752 F.2d 220, 228 (6th Cir. 1985).

[87] *The Supreme Court, 1986-Term—Leading Cases*, 101 HARV. L. REV. 119, 230 (1987).

[88] United States v. Attson, 900 F.2d 1427, 1430 (9th Cir. 1990) (emphasis in original).

[89] Ferguson v. City of Charleston, 532 U.S. 67, 77-78 (2001).

[90] Skinner v. Ry. Labor Executives' Ass'n, 489 U.S. 602, 620 (1989) (quoting Griffin v. Wisconsin, 483 U.S. 868, 873 (1987)).

enforcement purposes but for other reasons such as public safety[91] or ensuring the integrity of sensitive government positions,[92] then the doctrine will apply. If, however, the "primary purpose" or "immediate objective" was "to generate evidence *for law enforcement purposes*," then application of the special needs doctrine is not appropriate, and the government must adhere to general Fourth Amendment principles.[93] Again, the primary inquiry is the purpose of the search.

Some circuit courts of appeal have extended the special needs theory, holding that the Fourth Amendment does not apply (in contrast to a reduced standard of suspicion as with the special needs cases) unless the "conduct has as its purpose the intention to elicit a benefit for the government in either its investigative or administrative capacities."[94] In *United States v. Attson*, the Ninth Circuit held that the collection of blood by a government-employed physician, which was subsequently used by the police in a drunk driving prosecution, was not within the scope of Fourth Amendment protection.[95] The panel reasoned that the doctor drew the blood for medical purposes, not to further a governmental purpose in obtaining evidence against the defendant in its criminal investigation, so the Fourth Amendment did not apply.[96]

Applying these two theories to smart meters, a court would focus on the publicly owned utility's purpose in collecting the data. If it were for ordinary business purposes such as billing, informing the customer of its usage patterns, or aiding the utility in making the grid more energy-efficient, then it would not violate the Fourth Amendment. If, however, the public utility began aggregating data at the request of a law enforcement agency, with the purpose of aiding a criminal investigation or other administrative purpose, the Fourth Amendment would seemingly apply. As with private utilities, if the government requested that the public utility report any suspicious electricity usage, or created a program where certain data was regularly transmitted to the police, this might become investigatory and warrant Fourth Amendment protections. It appears law enforcement cannot evade Fourth Amendment restrictions by requesting a publicly owned utility to collect data for it.

Law enforcement might also request smart meter data under a public records theory. It is generally accepted that public records are not accorded Fourth Amendment protection.[97] Unless there is a state or federal statute prohibiting disclosure, "law enforcement access to state public records is unrestricted."[98] Thus the inquiry hinges on whether a document is a public record.

---

[91] *Id.*

[92] Nat'l Treasury Employees Union v. Von Raab, 489 U.S. 656, 670 (1989).

[93] *Ferguson*, 532 U.S. at 83 (emphasis in original).

[94] *See* United States v. Attson, 900 F.2d 1427, 1431 (9th Cir. 1990); Poe v. Leonard, 282 F.3d 123, 137 (2d Cir. 2002); United States v. Elliot, 676 F. Supp. 2d 431, 435-36 (D. Md. 2009).

[95] *Attson*, 900 F.2d at 1433.

[96] *Id.*

[97] *See* Nilson v. Layton City, 45 F.3d 369, 372 (10th Cir. 1995) ("Information readily available to the public is not protected by the constitutional right to privacy."); Doe v. City of New York, 15 F.3d 264, 268 (2d Cir. 1994) ("Certainly, there is no question that an individual cannot expect to have a constitutionally protected privacy interest in matters of public record."); United States v. Ellison, 462 F.3d 557, 562 (6th Cir. 2006) (accessing license plate number from computer database held not an intrusion of a constitutionally protected area, thus not a Fourth Amendment "search"); United States v. Baxter, 492 F.2d 150, 167 (9th Cir. 1973) (holding that Fourth Amendment protections do not extend to telephone company toll and billing records); *see also* Christopher Slobogin, *The Search and Seizure of Computers and Electronic Evidence: Transaction Surveillance by the Government*, 75 Miss. L. J. 139, 156 (2005).

[98] Slobogin, *supra* note 97.

Whether a person's utility records are public records differs from state to state.[99] Some states deem records of a municipally owned and operated electric utility as public records open for public inspection, while others have accorded these records statutory and constitutional protections.

In Florida, for example, records kept in connection with the operation of a city-operated utility are considered public records.[100] A similar policy applies in Georgia, where all records of a government agency, including utility records, must be open for inspection.[101] South Carolina, too, takes a similar approach.[102] It is not clear, however, from the reported cases whether these statutes permit access to personally identifiable information or simply operating records of the utility. Oklahoma is more explicit, permitting access to "records of the address, rate paid for services, charges, consumption rates, adjustments to the bill, reasons for adjustment, the name of the person that authorized the adjustment, and payment for each customer."[103] Oklahoma does protect some confidentiality, including "credit information, credit card numbers, telephone numbers, social security numbers, [and] bank account information for individual customers."[104] Other states, like Washington, specifically protect personally identifiable utility records. Washington does not require a showing of probable cause, but instead "a reasonable belief" that the record will help establish the customer committed a crime.[105] North Carolina likewise states that any "[b]illing information compiled and maintained by a city or county or other public entity providing utility services in connection with the ownership or operation of a public enterprise" is not a public record.[106]

---

[99] Because the focus of this report is federal law and the Fourth Amendment, a full treatment of state privacy law is beyond its scope.

[100] *In re* Public Records—Records of Municipally Operated Utility, Op. Att'y Gen. Fla. 74-35 (1974), *available at* http://www.myfloridalegal.com/ago.nsf/Opinions/B4AED736C2272860852566B30067371A; *see* FLA. STAT. §119.01(1) (2008) ("It is the policy of this state that all state, county, and municipal records are open for personal inspection by any person.").

[101] *See* GA. CODE ANN. §50-18-70(b) (2011); Op. Att'y Gen. Ga. 2000-4 (2000) (requiring personal utility records of certain public employees to be disclosed under public records law). Georgia defines a "public record" as "all documents, papers, letters, maps, books, tapes, photographs, computer based or generated information, or similar material prepared and maintained or received in the course of the operation of a public office or agency." GA. CODE ANN. §50-18-70(a).

[102] In South Carolina, public records include "information in or taken from any account, voucher, or contract dealing with the receipt or expenditure of public or other funds by public bodies." S.C. CODE ANN. §30-4-50 (2011). *See* Kelsey M. Swanson, *The Right to Know: An Approach to Gun Licenses and Public Access to Government Records*, 56 UCLA L. REV. 1579, 1601 (2009).

[103] OKLA. STAT. tit. 51, §24A.10 (2011).

[104] *Id.*

[105] WASH. REV. CODE §42.56.335 (2011). In Washington, the following rule applies to public utility districts and municipally owned electrical utilities:

> A law enforcement authority may not request inspection or copying of records of any person who belongs to a public utility district or a municipally owned electrical utility unless the authority provides the public utility district or municipally owned electrical utility with a written statement in which the authority states that it suspects that the particular person to whom the records pertain has committed a crime and the authority has a reasonable belief that the records could determine or help determine whether the suspicion might be true. Information obtained in violation of this section is inadmissible in any criminal proceeding.

WASH. REV. CODE §42.56.335. The Washington Supreme Court has raised this protection to state constitutional status in *In re* Personal Restraint of Maxfield, 133 Wash. 2d 332, 344 (1997).

[106] However, the North Carolina public records law declares that "[n]othing contained herein is intended to limit public disclosure by a city or county of bill information: ... that is necessary to assist law enforcement, public safety, fire (continued...)

Determining whether a utility is a state actor or whether smart meter data is a public record are merely threshold matters. A finding that an entity is a state actor or data is public does not foreclose law enforcement's ability to retrieve customer smart meter data, but instead activates the next step of Fourth Amendment analysis: whether the government invaded a reasonable expectation of privacy.

Reasonable Expectation of Privacy in Smart Meter Data

Under the modern conception of the Fourth Amendment, the government may not intrude into an area in which a person has an actual expectation of privacy that society would consider reasonable.[107] In the case of smart meter data, the government presumably seeks records in the custody of third-party utilities on the energy use at a specific home. However, a significant body of cases has refused to recognize constitutionally protected privacy interests in information provided by customers to businesses as part of their commercial relationships.[108] This theory, the third-party doctrine, permits police access to the telephone numbers a person dials[109] and to a person's bank documents,[110] free from Fourth Amendment constraints.

There are two relevant differences, however, between smart meters and the traditional third-party cases that may warrant a shift in approach. First is the possible judicial unease with the notion that advancement of technology threatens to erode further the constitutional protection of privacy.[111] From that perspective, as technology progresses, society faces an ever-increasing risk that an individual's activities will be monitored by the government. This is coupled with the concern that the breadth and granularity of personal information that new technology affords provide a far more intimate picture of an individual than the more limited snapshots available through prior technologies. Do the richness and scope of new information technologies warrant increased constitutional scrutiny?

Second, smart meters can convey information about the activities that occur inside the home, an area singled out for specific textual protection in the Fourth Amendment and one deeply ingrained in Anglo-Saxon law.[112] Even when the Court declared that "the Fourth Amendment protects people, not places,"[113] ostensibly shifting away from a property-based conception of the Fourth Amendment, it has still carved out special protections for the home.[114] However, concomitant with the increased use of technology in our private lives is increased exposure of our private activities, including those conducted in the home. Commonly, we share more personal

---

(...continued)

protection, rescue, emergency management, or judicial officers in the performance of their duties." N.C. GEN. STAT. §132-1.1(c)(3).

[107] Katz v. United States, 389 U.S. 347, 361 (1967) (Harlan, J., concurring).

[108] *See* Smith v. Maryland, 442 U.S. 735 (1979).

[109] *Id.*

[110] United States v. Miller, 425 U.S. 435 (1976).

[111] Kyllo v. United States, 533 U.S. 27, 33-4 (2001) ("It would be foolish to contend that the degree of privacy secured to citizens by the Fourth Amendment has been entirely unaffected by the advance of technology.").

[112] *See* Entick v. Carrington, 19 How. St. Tr. 1029 (C.P. 1765).

[113] Katz v. United States, 389 U.S. 347, 351 (1967).

[114] *See* Orin S. Kerr, *The Fourth Amendment and New Technologies: Constitutional Myths and the Case for Caution*, 102 MICH. L. REV. 801, 809-10 (2004) [hereinafter Kerr, *Fourth Amendment and New Technologies*].

information, even as our concerns grow that more individuals, businesses, and others can glean more information about our personal lives as a matter of course. As with technology generally, does the fact that more of our lives are becoming "public" call for lesser or greater constitutional protection, and how does a "reasonable expectation"-based model continue to apply in a technologically intensive society?

This subpart will first look at the third-party doctrine as it is commonly conceived by the courts. Then it will discuss whether there are sufficient differences between the use of smart meters and traditional third-party cases to counsel against its application.

Third-Party Doctrine

Traditionally, there has been no Fourth Amendment protection for information a consumer gives to business as part of their business dealings.[115] This doctrine dates back to the secret agent cases, in which any words uttered to another person, including a government agent or informant, were not covered by the Fourth Amendment.[116] It was later extended to business records, giving police access to documents such as telephone records,[117] bank records,[118] motel registration records,[119] and cell phone records.[120] The Supreme Court has reasoned that the customers assume the risk that the information could be handed over to government authorities,[121] and also that they consent to such access.[122] Some lower courts have applied this theory to traditional analog utility meters.[123] This section discusses the possible application of the third-party doctrine to smart meters.

In *Miller v. United States*, agents of the Bureau of Alcohol, Tobacco, and Firearms (ATF) subpoenaed several banks for records pertaining to the defendant, including copies of the defendant's checks, deposit slips, and financial statements.[124] The defendant moved to suppress the records at trial, arguing that a warrantless retrieval of the bank records (his "private papers")[125] was an intrusion into an area protected by the Fourth Amendment. The Court

---

[115] Orin S. Kerr, *The Case for a Third-Party Doctrine*, 107 MICH. L. REV. 561, 563 (2009) [hereinafter Kerr, *Third-Party Doctrine*]. While the third-party doctrine has supporters like Professor Kerr, this group is overshadowed by its vocal detractors. Professor LaFave described its underpinnings as "dead wrong" and that the "Court's woefully inadequate reasoning does great violence to the theory of Fourth Amendment protection which the Court developed in *Katz*." LAFAVE, *supra* note 78, §2.7(c). Justice Sotomayor lent credence to this sentiment in *United States v. Jones*, where she posited that it "may be necessary to reconsider the premise that an individual has no reasonable expectation of privacy in information voluntarily disclosed to third parties." United States v. Jones, 565 U.S. ___, 5 (Sotomayor, J., concurring in the judgment and the opinion).

[116] United States v. White, 401 U.S. 745, 750 (1971) (holding that the Fourth Amendment "affords no protection to a wrongdoer's misplaced belief that a person to whom he voluntarily confides his wrongdoing will not reveal it.") (internal quotation marks omitted).

[117] Smith v. Maryland, 442 U.S. 735 (1979).

[118] United States v. Miller, 425 U.S. 435 (1976).

[119] United States v. Willis, 759 F.2d 1486, 1498 (11th Cir. 1985).

[120] United States v. Hynson, No. 05-576, 2007 WL 2692327, at *6 (E.D. Pa. Sept. 11, 2007).

[121] *Smith*, 442 U.S. at 744.

[122] Kerr, *Third-Party Doctrine*, *supra* note 115.

[123] United States v. McIntyre, 646 F.3d 1107 (8th Cir. 2011).

[124] *Miller*, 425 U.S. at 437-438.

[125] Brief for Respondent at 4, *Miller*, 425 U.S. 435 (No. 74-1179), 1975 WL 173642, at *4 ("The Fourth Amendment is historically rooted in a concern for control over personal and private information in the face of governmental demands (continued...)

---

disagreed, broadly declaring "the Fourth Amendment does not prohibit the obtaining of information revealed to a third-party and conveyed by him to Government authorities, even if it is revealed on the assumption that it will be used only for a limited purpose and the confidence placed in the third-party will not be betrayed."[126] The Court further noted that "the depositor takes the risk, in revealing his affairs to another, that the information will be conveyed by that person to the Government."[127]

Three years later, the Court extended the third-party doctrine to outgoing numbers dialed from a person's telephone.[128] In *Smith v. Maryland*, the defendant robbed a woman and began making obscene phone calls to her.[129] Suspecting Smith placed the calls, the police used a pen register to track the telephone numbers dialed from his phone.[130] The police failed to obtain a warrant or subpoena before installing the pen register.[131] The register revealed that Smith was in fact making the phone calls to the woman. In denying Smith's motion to suppress, the Court relied on the third-party doctrine, stating that "this Court consistently has held that a person has no legitimate expectation of privacy in information he voluntarily turns over to third parties."[132] As applied to the telephone context, the Court found that "[w]hen he used his phone, [Smith] voluntarily conveyed numerical information to the telephone company and 'exposed' that information to its equipment in the ordinary course of business. In so doing, [Smith] assumed the risk that the company would reveal to police the numbers he dialed."[133]

Traditionally, utility records have been handled similarly to bank records and telephone records. Several lower federal courts have held that customers do not have a reasonable expectation of privacy in their utility records, thereby permitting warrantless access to these records. In *United States v. Starkweather*, the Ninth Circuit held that a person does not have a reasonable expectation of privacy in his utility records.[134] The panel reasoned that (1) these records were no different from phone records, and thus did not justify a different constitutional result; and (2) the public was aware that such records were regularly maintained, thereby negating any expectation of privacy.[135] The Eighth Circuit has also upheld warrantless police access to utility records in *United States v. McIntyre*.[136] The Eighth Circuit panel distinguished *Kyllo*, declaring that the means of obtaining the information in *Kyllo* (a thermal-imaging device) was significantly more intrusive than simply subpoenaing the records from the utility company.[137] The court held that "the means to obtaining the information is legally significant."[138] Likewise, the court in *United*

(...continued)

for access and use.") (citing Entick v. Carrington, 19 How. St. Tr. 1029 (C.P. 1765)).

[126] *Miller*, 425 U.S. at 443.

[127] *Id.*

[128] Smith v. Maryland, 442 U.S. 735 (1979).

[129] *Id.* at 737.

[130] *Id.*

[131] *Id.*

[132] *Id.* at 743-44.

[133] *Id.* at 744.

[134] United States v. Starkweather, No. 91-30354, 1992 WL 204005, at *2 (9th Cir. Aug. 24, 1992).

[135] *Id.*

[136] United States v. McIntyre, 646 F.3d 1107 (8th Cir. 2011).

[137] *Id.* at 1111.

[138] *Id.*

States v. Hamilton held that the means of obtaining power records from a third-party by way of administrative subpoena as opposed to "intrusion on the home by 'sense enhancing technology'" is "legally significant," removing this type of situation from the *Kyllo*-home privacy line of cases into the *Miller*-third-party line.[139]

It is difficult to predict whether a court would extend this traditional third-party analysis to smart meters. The courts may seek to ensure the predictability and stability of the third-party doctrine generally and administration of utility services specifically, thus requiring a bright-line rule for all third-party circumstances.[140] There is an advantage to a rule that is easy to apply, that allows utilities to better govern their affairs, and does not permit "savvy wrongdoers [to] use third-party services in a tactical way to enshroud the entirety of their crimes in zones of Fourth Amendment protection."[141] However, there are three overarching considerations embodied in the use of smart meters that might weigh against the application of traditional third-party analysis. These include (a) a person's expectation of privacy while at home; (b) the breadth and granularity of private information conveyed by smart meters; (c) the lack of a voluntary assumption of the risk or consent to release of this data.

Privacy in the Home

The location of the search mattered little in the traditional third-party cases, but it may take on constitutional significance with smart meters.[142] In the case of smart meters, the information is generated in the home, an area accorded specific textual protection in the Fourth Amendment, and one the Supreme Court has persistently safeguarded.[143] In no uncertain terms the Court has asserted that "[a]t the very core [of the Fourth Amendment] stands the right of a man to retreat into his own home and there be free from unreasonable government intrusion."[144] Even as technology advances—whether a tracking or thermal-imaging device or something new—the Court has maintained this bulwark. Because of the significance of the home, access to smart

---

[139] United States v. Hamilton, 434 F. Supp. 2d 974, 980 (D. Or. 2006); Booker v. Dominion Va. Power, No. 3:09-759, 2010 U.S. Dist. LEXIS 44960, at *17 (E.D. Va. May 7, 2010); *see also* Samson v. State, 919 P.2d 171, 173 (Ala. App. 1996) (holding under state constitution that "utility records are maintained by the utility and do not constitute information in which society is prepared to recognize a reasonable expectation of privacy"); People v. Stanley, 86 Cal. Rptr. 2d 89, 94 (Cal. App. 1999) (same).

[140] *See* Duncan Kennedy, *Form and Substance in Private Law Adjudication*, 89 HARV. L. REV. 1687, 1710 (1976).

[141] Kerr, *Third-Party Doctrine*, *supra* note 115, at 564.

[142] In *Smith*, the "site of the call was immaterial for purposes of analysis" of that case. Smith v. Maryland, 442 U.S. 735, 743 (1979). Whether a person dials a telephone number from his home, a telephone booth, or any other location does not alter the nature of the activity, and thus does not affect the Fourth Amendment analysis. The privacy interests implicated are the same no matter where the call is placed. The same theory applies to bank records. It matters not where someone writes a check, or fills out a deposit slip—the privacy interest is the same.

[143] Payton v. New York, 445 U.S. 573, 589 ("The Fourth Amendment protects the individual's privacy in a variety of settings. In none is the zone of privacy more clearly defined than when bounded by the unambiguous physical dimensions of an individual's home—a zone that finds its roots in clear and specific constitutional terms: 'The right of the people to be secure in their ... houses ... shall not be violated.'") (quoting U.S. CONST. amend IV); Minnesota v. Carter, 525 U.S. 83, 99 (1998) (Kennedy, J., concurring) ("[I]t is beyond dispute that the home is entitled to special protection as the center of the private lives of our people. Security of the home must be guarded by law in a world where privacy is diminished by enhanced surveillance and sophisticated communication systems.").

[144] Silverman v. United States, 365 U.S. 505, 511 (1961).

---

meter data may prompt a doctrinal shift away from the third-party doctrine. Several home privacy cases shed light on this possible approach.[145]

In *Kyllo v. United States*, the Court had to decide whether the use of a thermal-imaging device from the outside of a home that detected the amount of heat coming from inside the home was a violation of the Fourth Amendment.[146] In *Kyllo*, an agent of the Department of the Interior suspected Danny Kyllo was growing marijuana in his home with the use of high-intensity lamps.[147] The agent used a thermal imager to scan the outside of Kyllo's apartment to determine if he was using these "grow" lamps.[148] Thermal imagers can detect energy emitting from the outside surface of an object.[149] When scanning the home, the thermal imager produced an image with various shades of black, white, or gray—the shades darker or lighter depending on the warmth of the area being scanned.[150] From the passenger seat of his car, the agent scanned Kyllo's home for several minutes.[151] From his scan, he determined that the area over the garage and one side of his home were relatively hot compared to neighboring homes.[152] Based on utility bills, informant tips, and the results of thermal imaging, the agents obtained a warrant to search Kyllo's home.[153] As suspected, inside the home the agents found a marijuana growing operation, including over 100 plants.[154]

Justice Scalia first posited that "with very few exceptions, the question whether a warrantless search of the home is reasonable must be answered no."[155] Searches of the home were historically analyzed under the common law doctrine of trespass,[156] but during the mid-20th century the Court instead anchored the Fourth Amendment to a conception of privacy.[157] While this test may be difficult to apply in the context of automobiles, telephone booths, or other public areas, it is made easier when concerning the home:

> In the case of the search of the interior of homes—the prototypical and hence most commonly litigated area of protected privacy—there is a ready criterion, with deep roots in the common law, of the minimal expectation of privacy that *exists*, and that is acknowledged

---

[145] In April 2012, the Supreme Court will hear oral arguments in its most recent home privacy case, Jardines v. Florida, 73 So. 3d 34 (Fla. 2011), *cert granted*, 2012 U.S. LEXIS 7 (Jan. 6, 2012) (No. 11-564), where it will decide whether a drug sniff at the front door of a suspect's house by a trained narcotics dog is a Fourth Amendment search requiring probable cause. This case should shed further light on the parameters of privacy surrounding the home.

[146] Kyllo v. United States, 533 U.S. 27, 29 (2001).

[147] *Id.*

[148] *Id.*

[149] *Id.*

[150] *Id.* at 29-30.

[151] *Id.* at 30.

[152] *Id.*

[153] *Id.*

[154] *Id.* The Ninth Circuit held that Kyllo had not exhibited a subjective expectation of privacy in the home because he did not attempt to prevent the heat emitting from the lamps from escaping his home. United States v. Kyllo, 190 F.3d 1041, 1046 (9th Cir. 1999). Further, the panel held that even if he had a subjective expectation of privacy, it was not a reasonable one since the imager "did not expose any intimate details of Kyllo's life." *Id.* at 1047.

[155] *Kyllo*, 533 U.S. at 31.

[156] *See* Olmstead v. United States, 277 U.S. 438 (1928).

[157] Katz v. United States, 389 U.S. 347, 361 (1967) (Harlan, J., concurring). The modern formulation of the reasonable expectation of privacy test derives not from the majority opinion but from Justice Harlan's concurrence.

---

to be reasonable. To withdraw protection of this minimum expectation would be to permit police technology to erode the privacy guaranteed by the Fourth Amendment.[158]

The Court ultimately held that "obtaining by sense-enhancing technology any information regarding the interior of the home that could not otherwise have been obtained without physical intrusion into a constitutionally protected area constitutes a search—at least where (as here) the technology in question is not in general public use."[159] *Kyllo* affirmed the notion that "an expectation of privacy in activities taking place inside the home is presumptively reasonable."[160]

The Court also protected home privacy by prohibiting the monitoring of the location of a beeper while inside a residence.[161] In *United States v. Karo*, with the consent of a government informant the police attached a beeper to the false bottom of a can of ether, which was sold to Karo.[162] The can of ether was transported between several residences and storage facilities.[163] The police used the beeper to monitor the location of the can several times while it was located inside of the residences.[164] The Court was asked to determine "whether the monitoring of a beeper in a private residence, a location not open to visual surveillance, violates Fourth Amendment rights of those who have a justifiable interest in the privacy of the residence."[165] The Court answered in the affirmative.

The Court reiterated the long-standing notion that "private residences are places in which the individual normally expects privacy free of governmental intrusion not authorized by a warrant, and that expectation is plainly one that society is prepared to recognize as justifiable."[166] Unless there are exigent circumstances, "searches and seizures inside a home without a warrant are presumptively unreasonable...."[167] The Court ultimately held that the warrantless monitoring of the beeper in the home was a Fourth Amendment violation.[168]

Kyllo and *Karo* demonstrate that the Supreme Court "has defended the home as a sacred site at the 'core of the Fourth Amendment.'"[169] Although neither the Supreme Court nor any lower federal court has ruled on the use of smart meters, a few propositions can be deduced from *Kyllo* and *Karo* bearing on this question.

Because smart meters allow law enforcement to access information regarding intimate details occurring inside the home, a highly invasive investigation that could not otherwise be performed without intrusion into the home, a court may require a warrant to access this data. In *Kyllo*, the

---

[158] *Kyllo*, 533 U.S. at 34.

[159] *Id.* (internal quotation marks omitted).

[160] Lerner & Mulligan, *supra* note 60, ¶ 18.

[161] United States v. Karo, 468 U.S. 705 (1984).

[162] *Id.* at 708.

[163] *Id.*

[164] *Id.* at 709-10.

[165] *Id.*

[166] *Id.* at 714.

[167] *Id.* at 714-15.

[168] *Id.* at 718.

[169] Stephanie M. Stern, *The Inviolate Home: Housing Exceptionalism in the Fourth Amendment*, 95 CORNELL L. REV. 905, 913 (2010) (citing Wilson v. Layne, 526 U.S. 603, 612 (1999)).

police merely obtained the relative temperatures of a house,[170] and in *Karo* the police only generally located the beeper in the house.[171] Although this information was limited, the Court nonetheless prohibited such investigatory techniques. Smart meters have the potential to produce significantly more information than that derived in *Kyllo* and *Karo*, including what individual appliances we are using; whether our house is empty or occupied; and when we take our daily shower or bath.[172] Further, a look at **Figure 1**, *supra*, makes it clear that this level of information is much more intimate than prior technologies used by law enforcement. This depth of intrusion suggests that customers may have a reasonable expectation of privacy in smart meter data.

There is also a question whether smart meters are in "general public use." (The police must use technology not in general public use for *Kyllo* to apply.)[173] Unfortunately, the Court provided no criterion for making this determination.[174] Several courts applying this test have held that night vision goggles were in general public use.[175] One federal district court reasoned that the goggles were regularly used by the military and police and could be found on the Internet, so were considered in general public use.[176] In 2009, the Department of Energy estimated that 4.75% of all electric meters were smart meters.[177] The department projects that by 2012 approximately 52 million more meters will be installed.[178] With little guidance on this issue, it is uncertain whether this jump in numbers would elevate smart meters into the general public use category.

The means by which data is gathered also differentiates the thermal-imaging in *Kyllo* from smart meters. In *Kyllo*, the police independently gathered the information using the thermal imager; an agent went outside Kyllo's house and used the thermal imager himself.[179] With smart meters, the utility company compiles the information and the police subpoena the company for the data. This difference in means was material in one lower court analyzing access to traditional utility data.[180] It is not clear whether this difference advises against application of *Kyllo* here.

Mosaic and Dragnet Theories

The second factor guiding against the application of the third-party doctrine is composed of two interconnected theories: the mosaic and dragnet theories. The mosaic theory is grounded in the idea that surveillance of the whole of one's activities over a prolonged period is substantially

---

[170] United States v. Kyllo, 533 U.S. 27, 30 (2001).

[171] *Karo*, 468 U.S. at 705, 709-10.

[172] NIST PRIVACY REPORT, *supra* note 11, at 14 & n.35. It is unclear whether the specificity of the data from the smart meter will directly affect the constitutional analysis. *Kyllo*, 533 U.S. at 37 ("The *Fourth Amendment's* protection of the home has never been tied to measurement of the quality or quantity of information obtained."). With that said, the NIST report maintains that sufficient information about the activities inside of the home are presented to implicate a *Kyllo*, home search analysis.

[173] *Kyllo*, 533 U.S. at 34.

[174] *See* Douglas Adkins, *The Supreme Court Announces a Fourth Amendment "General Public Use" Standard for Emerging Technologies but Fails to Define It:* Kyllo v. United States, 27 DAYTON L. REV. 245 (2002).

[175] *See* United States v. Dellas, 355 F. Supp. 2d 1095, 1107 (N.D. Cal. 2005).

[176] United States v. Vela, 486 F. Supp. 2d 587, 590 (W.D. Tex. 2005).

[177] DEP'T OF ENERGY, SMART GRID SYSTEM REPORT vi (2009), *available at* http://energy.gov/sites/prod/files/oeprod/DocumentsandMedia/SGSRMain_090707_lowres.pdf.

[178] *Id.*

[179] United States v. Kyllo, 533 U.S. 27, 29 (2001).

[180] United States v. McIntyre, 646 F.3d 1107, 1111-12 (8th Cir. 2011).

---

more invasive than a look at each item in isolation.[181] In the case of smart meters, this is the difference between knowing a person's monthly energy usage, and being able to discern a person's daily activities with considerable accuracy. This theory intersects with dragnet-styled law enforcement techniques in which the police cast a wide surveillance net, taking in a wealth of personal information with the goal of finding criminal activity among the stream of data.

Although the Supreme Court has never formally adopted the mosaic theory, there seems to be a ready-made majority potentially willing to consider it.[182] In *United States v. Jones*, the police used a GPS tracking device to track Jones's movements for almost a month.[183] The majority, led by Justice Scalia, held that attaching a GPS device on a vehicle for the purpose of collecting information constituted a "search" under the Fourth Amendment.[184] The physical intrusion, rather than a *Katz*-type invasion of privacy, was the lynchpin of the decision.[185] Justices Alito and Sotomayor both agreed that this was a search, but on different grounds. Both discussed an adaptation of the mosaic theory as prohibiting police from tracking a person for an extended period of time. Justice Alito, joined by Justices Breyer, Ginsburg, and Kagan, assumed that a short-term search would not violate the Fourth Amendment, but that "the use of longer term GPS monitoring in investigations of most offenses impinges on expectations of privacy."[186] Likewise, Justice Sotomayor agreed with this "incisive" observation, noting that "GPS monitoring generates a precise, comprehensive record of a person's public movements that reflects a wealth of detail about familial, political, professional, religious, and sexual associations."[187] Both of these comments closely mirror those of the opinion below, which relied on the mosaic theory: "A person who knows all of another's travels can deduce whether he is a weekly church goer, a heavy drinker, a regular at the gym, an unfaithful husband, an outpatient receiving medical treatment, an associate of particular individuals or political groups—and not just one such fact about a person, but all such facts."[188]

Although the *Jones* majority did not embrace the mosaic theory, the concurrences demonstrate that five justices are flirting with the idea. These arguments resemble those made against the unfettered use of smart meter data. With smart meters, police would have a rich source of personal data that reveals far more about a person than traditional analog meters. Understanding a person's daily activities, including what appliances he is using, is a far leap from knowing his monthly energy usage. This is the difference between knowing about a single trip a person took and monitoring his movements over a month-long period. The breadth and granularity of the smart meter data may be seen as warranting application of the mosaic theory and may perhaps find receptive ears on the Court.

Additionally, the dragnet theory may apply to collection of energy usage data. This theory states that surveillance normally permitted under the Fourth Amendment—such as monitoring a person's movements on a public street—becomes an impermissible invasion of privacy when

---

[181] *See* Cent. Intelligence Agency v. Sims, 471 U.S. 159, 178 (1985).

[182] *See* Orin Kerr, VOLOKH CONSPIRACY, What's the Status of the Mosaic Theory After Jones?, http://volokh.com/2012/01/23/whats-the-status-of-the-mosaic-theory-after-jones/.

[183] United States v. Jones, 565 U.S. ___, 2 (2012).

[184] *Id.* at 3.

[185] *Id.* at 4.

[186] *Id.* at 13 (Alito, J., concurring in the judgment).

[187] *Id.* at 3 (Sotomayor, J., concurring in the judgment and the opinion).

[188] United States v. Maynard, 615 F.3d 544, 562 (D.C. Cir. 2010).

conducted on a prolonged, 24-hour basis.[189] "If such dragnet-type law enforcement practices as respondent envisions should eventually occur," Justice Rehnquist asserted earlier in *United States v. Knotts*, "there will be time enough then to determine whether different constitutional principles may be applicable."[190] Twenty-four hour access to our intimate daily activities, including what appliances we use, when we take our daily shower or bath, eat, and sleep, may push smart meters into the dragnet category.

Coinciding with the mosaic and dragnet theories is the difference in sophistication and the quantity of the data revealed between traditional third-party cases and smart meters. Comparing *Smith* with *Katz* provides insight into this distinction. Pen registers, as used in *Smith*, have "limited capabilities"—they can only record the numbers dialed from a phone.[191] In comparison, in *Katz* the police listened to the contents of Katz's phone call—the actual words spoken.[192] In noting this distinction, it seems the *Smith* Court, in permitting the use of pen registers, intentionally limited its holding to the discrete set of data conveyed—the telephone numbers dialed. Smart meters, to the contrary, have the potential to collect and aggregate precise detail about the activities inside the home. It is more than one packet of data, but reveals minute-by-minute activity, something far more revealing, and arguably more like *Katz* than *Smith*.

Assumption of the Risk—Consent

The third difference between traditional third-party cases and smart meters is the nature of services involved and whether the customer actually assumes the risk or consents to this information being shared with others. Assumption of the risk and consent are the two leading theories supporting the third-party doctrine. In *United States v. Miller*, the customer "assumed the risk" that the bank would turn over the bank records to government authorities.[193] That was a risk he took in doing business with the bank. As to the consent theory, one commentator asked and answered the question as follows: "When does a person's choice to disclose information to a third-party constitute consent to a search? So long as a person knows that they are disclosing information to a third-party, their choice to do so is voluntary and the consent valid."[194]

With banking or telephone services, a customer has the option of transferring his business to another bank or another telephone carrier.[195] To the contrary, because electric utilities are essentially monopolies, the customer cannot simply switch services. The only way to avoid the recordation of his electric usage is to terminate his utility service altogether, an impracticable option in modern society. As one state court has noted:

> Electricity, even more than telephone service, is a "necessary component" of modern life, pervading every aspect of an individual's business and personal life: it heats our homes,

---

[189] *Id.* at 558.

[190] United States v. Knotts, 460 U.S. 276, 283-84 (1983). Because this statement was not essential to the holding, it was dictum: persuasive, but not binding.

[191] *Smith*, 442 U.S at 741 (citing Katz v. United States, 389 U.S. 347 (1967)).

[192] *Katz*, 389 U.S. at 348.

[193] *Smith*, 442 U.S. at 744 (citing United States v. Miller, 425 U.S. 435 (1976)).

[194] Kerr, *Third-Party Doctrine, supra* note 115, at 588.

[195] *Contra Smith*, 442 U.S. at 750 (Marshall, J., dissenting) ("[U]nless a person is prepared to forgo use of what for many has become a personal or professional necessity, he cannot help but accept the risk of surveillance. It is idle to speak of "assuming" the risk in contexts where, as a practical matter, individuals have no realistic alternative.").

> powers our appliances, and lights our nights. A requirement of receiving this service is the disclosure to the power company (and in this case an agent of the state) of one's identity and the amount of electricity being used. The nature of electrical service requires the disclosure of this information, but that disclosure is only for the limited business purpose of obtaining the service.[196]

It is not clear whether assumption of the risk or consent should apply to smart meters. It is reasonable to assume that customers understand utility companies must collect usage data to bill the customer for that usage. Customers receive their statement each month demonstrating this fact. However, most customers are probably not familiar with the sophistication of smart meters and the detailed data sets that can be derived from them. Even if customers are aware their utility usage can be recorded in sub-fifteen minute intervals, a reasonable customer would probably be surprised, if not shocked, to know that data from smart meters can potentially be used to pinpoint the usage of specific appliances. If knowledge of the sophistication of the data is a prerequisite to assumption of the risk or consent, it is difficult to say whether a reasonable customer would understand the privacy implications with this new technology.[197]

Because smart meters are an emerging technology not yet judicially tested, it is difficult to conclude with certainty how they would be handled under the Fourth Amendment. Further, beyond the possible constitutional implications of smart meters, federal communication and privacy statutes may also apply. As noted by Professor Kerr, "in recent decades, legislative privacy rules governing new technologies have proven roughly as privacy protective, and quite often more protective than, parallel Fourth Amendment rules."[198]

Statutory Protection of Smart Meter Data

This section discusses federal statutory protections that may be applicable to the contents of communications sent by a smart meter, independent of the Fourth Amendment, while they are either stored within the smart meter prior to transmission, during transmission, or after they have been delivered to the utility. Three federal laws, the Electronic Communications Privacy Act (ECPA),[199] the Stored Communications Act (SCA),[200] and the Computer Fraud and Abuse Act (CFAA)[201] may be applicable to these situations and are discussed in more detail below.

---

[196] *In re* Restraint of Maxfield, 133 Wn.2d 332, 341 (Wash. 1997); *see also* Balough, *supra* note 63, at 185.

[197] *Cf.* United States v. Warshak, 631 F.3d 266, 288 (6th Cir. 2010) ("*Miller* involved simple business records, as opposed to the potentially unlimited variety of 'confidential communications' at issue here.").

[198] Kerr, *Fourth Amendment and New Technologies, supra* note 114, at 806.

[199] For more detailed information on ECPA, *see* CRS Report R41733, *Privacy: An Overview of the Electronic Communications Privacy Act*, by Charles Doyle.

[200] For a more detailed discussion of the SCA, *see* CRS Report R41733, *Privacy: An Overview of the Electronic Communications Privacy Act*, by Charles Doyle.

[201] For more detailed information on the CFAA, *see* CRS Report 97-1025, *Cybercrime: An Overview of the Federal Computer Fraud and Abuse Statute and Related Federal Criminal Laws*, by Charles Doyle.

The Electronic Communications Privacy Act (ECPA)

ECPA, enacted in 1986, "addresses the interception of wire, oral and electronic communications."[202] The statute defines electronic communications as "any transfer of signs, signals, writing, images, sounds, data, or intelligence of any nature transmitted in whole or in part by a wire, radio, electromagnetic, photoelectronic or photooptical system that affects interstate or foreign commerce...."[203] Based on the description of the smart meter network provided above,[204] the envisioned transmission of customers' energy usage data by smart meters would seem to fall squarely within the definition of electronic communications under ECPA.

ECPA generally prohibits the interception of electronic communications, but also provides a mechanism for government entities to conduct such surveillance, and a number of other exceptions.[205] Additionally, the statute provides that interception under the procedures and exceptions set forth in ECPA, or pursuant to the Foreign Intelligence Surveillance Act, are the exclusive means for intercepting electronic communications.[206] The unlawful interception of electronic communications in violation of ECPA is generally punishable by imprisonment for not more than five years and/or a fine of not more than $250,000 for individuals and not more than $500,000 for organizations.[207]

Of particular relevance to the immediate discussion is the fact that ECPA permits interception of an electronic communication where a party to the communication has consented to such interception.[208] In the context of a smart meter network that is the subject of this report, it appears that the utility would be a party to all of the communication sent by the smart meters, since it is primarily receiving that information for its own billing purposes. Therefore, if the utility consents to law enforcement's interception of the traffic which is addressed to it, that surveillance would not appear to violate the prohibitions in ECPA.

ECPA also provides a procedural mechanism for law enforcement to conduct surveillance activities for investigative purposes without the consent of any party to the communication. The statute limits the types of criminal cases in which electronic surveillance may be used[209] and requires court orders authorizing electronic surveillance to be supported by probable cause to believe that the target is engaged in criminal activities, that normal investigative techniques are

---

[202] S.Rept. 99-541 at 3.

[203] 18 U.S.C. §2510(12).

[204] *See supra* note 47 and accompanying text (noting that smart meters may use a variety of communications technologies, including fiber optics, wireless networks, satellite, and broadband over power line).

[205] 18 U.S.C. §2516. Exceptions cover things such as interception with the consent of a party to the communication and interception by communication service providers as an incident to providing service.

[206] 18 U.S.C. §2511(2)(f). FISA defines electronic surveillance to include more than the interception of wire, oral, or electronic communications, 50 U.S.C. §1801(f), but places limitations on its definition based upon the location or identity of some or all of the parties to the communications involved.

[207] "Except as provided in (b) of this subsection or in subsection (5), whoever violates subsection (1) of this section shall be fined under this title or imprisoned not more than five years, or both." 18 U.S.C. §2511(4)(a).

[208] 18 U.S.C. §2511(2)(c).

[209] The list of covered criminal provisions can be found at 18 U.S.C. §2516(1), and includes offenses such as violence at international airports; animal enterprise terrorism; arson; bribery of public officials and witnesses; unlawful use of explosives; fraud by wire, radio, or television; terrorist attacks against mass transportation; sexual exploitation of children; narcotics production and trafficking; and many others.

insufficient, and that the facilities that are the subject of surveillance will be used by the target.[210] It also limits the use and dissemination of information intercepted.[211] In addition, when an interception order expires, authorities must notify those whose communications have been intercepted.[212] Law enforcement may also conduct electronic surveillance when acting in an emergency situation pending issuance of a court order.[213]

The government may also conduct electronic surveillance under the authority of the Foreign Intelligence Surveillance Act (FISA). FISA governs the gathering of information about foreign powers, including international terrorist organizations, and agents of foreign powers.[214] Although it is often discussed in relation to the prevention of terrorism, it applies to the gathering of foreign intelligence information for other purposes.[215] Although some exceptions apply, such as for emergency situations,[216] the government typically must obtain a court order, supported by probable cause, from the Foreign Intelligence Surveillance Court (FISC), a neutral judicial decision maker, in order to conduct electronic surveillance pursuant to FISA.[217]

The Stored Communications Act (SCA)

The SCA was enacted in 1986 as Title II of the Electronic Communications Privacy Act (ECPA),[218] to "address[] access to stored wire and electronic communications and transactional records."[219] The SCA prohibits unauthorized persons from accessing a facility through which an *electronic communication service* (ECS) is provided; or obtaining, altering, or preventing access to an electronic communication while it is in *electronic storage* in an ECS.[220] The SCA also limits the circumstances in which providers of ECS or a *remote computing service* (RCS) may disclose information that they carry or maintain.[221] The SCA also provides a mechanism by which law enforcement may compel the disclosure of stored communications.[222]

The terms "electronic communication service," "remote computing services," and "electronic storage" are all specifically defined by the SCA. As described above, the SCA applies only to providers of either an ECS or an RCS; stored communications held by other types of entities are not protected by the SCA. Therefore, in order to determine whether the SCA would protect stored information collected by a smart meter, this report will first examine whether a utility's deployment of a smart meter network falls within the definition of an ECS or an RCS and then

---

[210] 18 U.S.C. §§2516, 2518(3).

[211] 18 U.S.C. §2517.

[212] 18 U.S.C. §2518(8).

[213] 18 U.S.C. §2518(7).

[214] *See* 50 U.S.C. §1801(a) (definition of "foreign power").

[215] For example, it extends to the collection of information necessary for the conduct of foreign affairs. *See* 50 U.S.C. §1801(e) (definition of "foreign intelligence information").

[216] 50 U.S.C. §1805(e).

[217] 50 U.S.C. §§1801-1808. FISA authorizes electronic surveillance without a FISA order in specified instances involving communications between foreign powers. 50 U.S.C. §1802.

[218] P.L. 99-508.

[219] S.Rept. 99-541 at 3.

[220] 18 U.S.C. §2701(a). Unauthorized access includes exceeding an authorization to use the facility. *Id.*

[221] 18 U.S.C. §2702.

[222] 18 U.S.C. §2703.

discuss the protections and disclosure restrictions that might apply to any smart meter network that qualifies as an ECS or RCS.

Electronic Communication Services

An ECS is defined by the SCA as any service which provides users "the ability to send or receive wire or electronic communications."[223] The statute also defines an "electronic communication" as "any transfer of signs, signals, writing, images, sounds, data, or intelligence of any nature transmitted in whole or in part by a wire, radio, electromagnetic, photoelectronic or photooptical system that affects interstate or foreign commerce."[224] As described above, one of the essential functions of a smart meter would appear to be the capability to transmit consumer electricity usage data to the smart grid using a variety of communications technologies.[225] These transmissions would seem to fall neatly within the SCA's definition of an electronic communication. Therefore, whether a smart meter network would qualify as an ECS would likely depend on whether the deployed smart meters could be said to be providing this ability to users.

It is not clear whether it would be accurate to categorically describe smart meters as providing customers with "the ability to send or receive" communications. It could be argued that a utility customer would use the smart meter to transmit usage information to the utility, in the same way that the same customer uses a traditional meter to record household electricity usage over a billing period. However, the Ninth Circuit has suggested that an ECS should not include situations in which electronic communications are used only "as an incident to providing some other service, as is the case with a street-front shop that requires potential customers to speak into an intercom device before permitting entry, or a 'drive-thru' restaurant that allows customers to place orders via a two-way intercom located beside the drive-up lane."[226] On one hand, it may not be accurate to describe utility customers as users of smart meters at all, particularly if the deployment of such smart meters is intended principally for the benefit of the utility and does not change the experience of utility customers. On the other hand, some of the proposed uses of deployed smart meters may include using collected data for the benefit of the customers, for example by determining the energy efficiency of specific household appliances.[227] As a result, the ultimate classification of a particular smart meter network as an ECS may depend largely on the specific facts present, such as the manner in which it is marketed, or the ostensible purposes for which the transmissions are intended to be used.

If a smart meter network qualifies as an ECS, then transmissions containing smart meter data would be protected under the SCA only while such transmissions are in electronic storage, as that term is defined by the statute.[228] Therefore, one must first determine whether, and under what circumstances, the data collected by a smart meter network is in electronic storage in order to determine what protections apply.

---

[223] 18 U.S.C. §2510(15).

[224] 18 U.S.C. §2510(12). Wire communications are defined as communications containing the human voice and are not implicated here. 18 U.S.C. §2510(1).

[225] *See supra* note 47 and accompanying text.

[226] Company v. United States (*In re* United States), 349 F.3d 1132, 1141 (9th Cir. 2003) (holding that definition of ECS includes service that provides drivers with the ability to make phone calls from their car for directory assistance, driving directions, or roadside assistance because those activities are intrinsically communicative).

[227] *See supra* note 8.

[228] 18 U.S.C. §2701.

For purposes of the SCA, a communication is in electronic storage at an ECS if it is in temporary, intermediate storage incidental to electronic transmission or in storage for backup protection.[229] As applied to the smart meter network, data residing on the smart meter itself prior to being sent to the utility would appear to be in electronic storage, as such storage is likely temporary and undertaken solely in anticipation of some eventual transmission to the utility. In contrast, once the data has arrived at the utility and resides on its servers, it may no longer be in temporary or intermediate storage. However, some form of the communications may still be being held for backup purposes, and in such a case might be considered in electronic storage under the statute. To the extent that the data would be considered in electronic storage, either while on the meter or on the utility's computers, the data would appear to be subject to the SCA's provisions applicable to providers of ECS.

The SCA prohibits intentionally accessing without authorization, a facility through which an ECS is provided and obtaining, altering, or preventing access to an electronic communication while it is in electronic storage.[230] Criminal penalties for violating the SCA's prohibitions on unauthorized access start at imprisonment for not more than one year (not more than five years for a subsequent conviction) and/or a fine of not more than $100,000.[231] However, violations committed for malicious, mercenary, tortious or criminal purposes are subject to higher penalties and may be punished by imprisonment for not more than five years (not more than 10 years for a subsequent conviction) and/or a fine of not more than $250,000 (not more than $500,000 for organizations).[232] Victims of a violation of the SCA also have a civil cause of action for equitable relief, reasonable attorneys' fees and costs, and damages equal to the loss and gain associated with the offense but not less than $1,000.[233]

The SCA generally restricts the ability of providers of ECS to disclose the contents of communications in electronic storage, if the ECS is offering those services to the public.[234] However, the statute also permits certain disclosures to law enforcement. Such permitted disclosures by a provider of electronic communication services to law enforcement can be either voluntary or compelled. Normally, voluntary disclosure to law enforcement is authorized only if the contents of the communication were inadvertently obtained by the service provider and appear to pertain to the commission of a crime.[235] However, it should be noted that the utility in this case appears to be the intended recipient of all communications sent over the smart meter network, and the SCA's restrictions on disclosures of electronically stored information held by ECS or RCS providers may generally be overcome if an intended recipient of the communication consents to the disclosure.[236] Consequently, the utility may have more latitude to share communications in electronic storage with law enforcement than a traditional provider of ECS, such as a telephone company, would have.

---

[229] 18 U.S.C. §2510(17).

[230] 18 U.S.C. §2701(a). Unauthorized access includes exceeding an authorization to use the facility. *Id.*

[231] 18 U.S.C. §2701(b)(2).

[232] 18 U.S.C. §2701(b)(1).

[233] 18 U.S.C. §2707.

[234] 18 U.S.C. §2702(a)(1) ("a person or entity providing an electronic communication service to the public shall not knowingly divulge to any person or entity the contents of a communication while in electronic storage by that service").

[235] 18 U.S.C. §2702(b)(7).

[236] *See* 18 U.S.C. §2702(b)(3).

For purposes of compelled disclosures to law enforcement, the SCA distinguishes between recent communications and those that have been in electronic storage for more than 180 days. A search warrant is required to compel providers to disclose communications held in electronic storage for 180 days or less.[237] However, communications held for more than 180 days may be obtained by law enforcement through a warrant, subpoena, or a court order supported by specific and articulable facts sufficient to establish reasonable grounds to believe that the contents are relevant and material to an ongoing criminal investigation.[238] Customers whose communications have been disclosed are generally required to be given notice of such disclosure, but such disclosure may be delayed if notification might result in endangering the life or physical safety of an individual; flight from prosecution; destruction of or tampering with evidence; intimidation of potential witnesses; or otherwise seriously jeopardizing an investigation or unduly delaying a trial.[239]

Remote Computing Services

It is likely that the classification of a smart meter network as an RCS would similarly be fact-dependent. The SCA defines an RCS as a service in which computer storage or processing services by means of an ECS are provided to the public.[240] It is conceivable that the data collected by smart meters may in fact be stored or processed by the utility, but there is no indication that such storage or processing would be categorically provided as a service to the public, rather than solely for the utility's internal benefit.[241] If such service is not provided to the public, then it would likely be inaccurate to classify the smart meter network as an RCS. However, if one of the features of a particular smart meter deployment is to give customers the ability to store or process their usage data, then it would appear to qualify as an RCS.

For those smart meter networks which qualify as an RCS, the SCA generally protects the contents of electronically transmitted communications "carried or maintained on that service" for customers of the service. Disclosures of such information are generally prohibited,[242] but the SCA also provides a means for law enforcement to obtain access to the contents of such communications. The government may obtain a warrant supported by probable cause, or use a subpoena or a court order supported by specific and articulable facts sufficient to establish reasonable grounds to believe that the contents are relevant and material to an ongoing criminal investigation.[243] However, use of a subpoena or court order supported by specific and articulable facts also requires the government to give prior notice to the customer whose information is sought, unless particular circumstances warrant delayed notice.[244] RCS customers whose

---

[237] 18 U.S.C. §2703(a).

[238] 18 U.S.C. §2703(d). Some courts have held that this "reasonable grounds" standard is a less demanding standard than "probable cause." *See In re* Application of the United States, 620 F.3d 304, 313 (3d Cir. 2010) ("We also conclude that this [§2703(d)] standard is a lesser one than probable cause.").

[239] 18 U.S.C. §2705(a).

[240] 18 U.S.C. §2711(2).

[241] However, if some other service provided by the utility allows the data collected by a smart meter to be stored or manipulated for the benefit of the utility's customers, it is possible that this system would fall within the definition of an RCS.

[242] The SCA allows providers of an RCS to disclose stored communications with the consent of the subscriber of an RCS. 18 U.S.C. §2702(b)(3).

[243] 18 U.S.C. §2703(b)(1).

[244] 18 U.S.C. §2703(b)(1)(B).

communications have been disclosed in violation of the SCA may pursue a civil cause of action for equitable relief, reasonable attorneys' fees and costs, and damages equal to the loss and gain associated with the offense but not less than $1,000.[245]

The Computer Fraud and Abuse Act (CFAA)

The Computer Fraud and Abuse Act (CFAA) prohibits intentionally accessing and obtaining information from a computer used in or affecting interstate commerce, without authorization or in excess of a granted authorization.[246] The definition of a computer for purposes of the CFAA is "an electronic, magnetic, optical, electrochemical, or other high speed data processing device performing logical, arithmetic, or storage functions, and includes any data storage facility or communications facility directly related to or operating in conjunction with such device" excluding "an automated typewriter or typesetter, a portable hand held calculator, or other similar device...."[247]

The servers on a utility's network would likely fall squarely within the definition of a computer under the CFAA. Similarly, smart meters themselves also appear to meet the definition of a computer, insofar as they store customers' energy usage data and also perform logical operations by routing transmissions across the utility's network. Additionally, in light of the significant role that energy utilities play in the modern economy, the smart meter network would also likely be considered to have an effect on interstate commerce, even if they operate entirely within one state. Therefore, intentionally gaining access to the utility's servers or smart meters to obtain customer data would likely constitute a violation of the CFAA if done without the utility's authorization or in excess of an authorization granted by the utility.

The criminal penalties for violating the unauthorized access provisions of the CFAA have a three tier sentencing structure. Simple violations are punished as misdemeanors, imprisonment for not more than one year and/or a fine of not more than $100,000 ($200,000 for organizations).[248] At the next level, cases in which: "(i) the offense was committed for purposes of commercial advantage or private financial gain; (ii) the offense was committed in furtherance of any criminal or tortious act in violation of the Constitution or laws of the United States or of any State; or (iii) the value of the information obtained exceeds $5,000" may be punished by imprisonment for not more than five years and/or a fine of not more $250,000 ($500,000 for organizations).[249] The third tier is for repeat offenders whose punishment is increased to imprisonment of not more than 10 years and/or a fine of not more than $250,000 ($500,000 for organizations) for a second or subsequent conviction.[250]

---

[245] 18 U.S.C. §2707.

[246] 18 U.S.C. §1030(a)(2). For more detailed information on the CFAA, *see* CRS Report 97-1025, *Cybercrime: An Overview of the Federal Computer Fraud and Abuse Statute and Related Federal Criminal Laws*, by Charles Doyle.

[247] 18 U.S.C. §1030(e)(1).

[248] 18 U.S.C. §1030(c)(2)(A).

[249] 18 U.S.C. §1030(c)(2)(B).

[250] 18 U.S.C. §§1030(c), 3571.

The Federal Trade Commission Act (FTC Act)

Section 5 of the FTC Act prohibits "unfair or deceptive acts or practices in or affecting commerce"[251] and gives the Federal Trade Commission (FTC) jurisdiction to bring enforcement actions against "persons, partnerships, or corporations" that engage in these practices.[252] In the past, the FTC has used its authority under Section 5 to take action against businesses that violate their own privacy policies or that fail to adequately safeguard a consumer's personal information.[253] Although there do not appear to be any cases in which the FTC has taken action against an electric utility for failing to protect consumer smart meter data, the Commission would have authority to enforce Section 5 against a utility that fell within its statutory jurisdiction.

Covered Electric Utilities

This section considers whether the FTC would have Section 5 jurisdiction over each of the four types of electric utilities identified by the Energy Information Administration (EIA): investor-owned, publicly owned, federally owned, and cooperative.[254] It finds that the FTC clearly has jurisdiction over investor-owned utilities. It is unclear whether the Commission has jurisdiction over publicly owned utilities or federally owned utilities. The FTC could enforce Section 5 against for-profit electric cooperatives, and case law suggests that nonprofit electric cooperatives may also be subject to the act's requirements.

The FTC has jurisdiction to enforce Section 5 against "persons, partnerships, or corporations," with exceptions not applicable here.[255] Utilities that are "persons" or "partnerships" would be subject to the FTC's enforcement powers automatically,[256] as the statute does not provide any additional jurisdictional requirements for these entities. Most electric utilities, however, are organized as legal entities that would potentially fit within the definition of "corporation." The FTC Act states that, for the purposes of Section 5, the term "corporation":

> shall be deemed to include any company, trust, so-called Massachusetts trust, or association, incorporated or unincorporated, which is organized to carry on business for its own profit or that of its members, and has shares of capital or capital stock or certificates of interest, and any company, trust, so-called Massachusetts trust, or association, incorporated or unincorporated, without shares of capital or capital stock or certificates of interest, except partnerships, which is organized to carry on business for its own profit or that of its members.[257]

---

[251] 15 U.S.C. §45(a)(1).

[252] 15 U.S.C. §45(a)(2).

[253] *See* "Enforcement of Data Privacy and Security," *infra* p. 41; *see also* NIST PRIVACY REPORT, *supra* note 11, at 23 n.48.

[254] ENERGY INFO. ADMIN., ELECTRIC POWER INDUSTRY OVERVIEW (2007) [hereinafter EIA ELECTRIC POWER OVERVIEW], *available at* http://www.eia.gov/cneaf/electricity/page/prim2/toc2.html.

[255] 15 U.S.C. §45(a)(2).

[256] The FTC Act does not further define "persons" or "partnerships" or impose any additional jurisdictional requirements on these entities in the way that it does for "corporations." *See* 15 U.S.C. §44.

[257] 15 U.S.C. §44.

This definition, particularly in its use of the words "shall be deemed to include," suggests that a wide variety of legal entities could potentially constitute "corporations." Moreover, in *California Dental Ass'n v. FTC*, the Supreme Court remarked that the "FTC Act directs the Commission to prevent the *broad set of entities* under its jurisdiction" from violating Section 5.[258] In that case, the Court found that the term "corporation" also included *nonprofit* entities, so long as they imparted significant economic benefit to their members.[259] Thus, as the Court's opinion demonstrates, the key question when determining whether an entity is a "corporation" for the purposes of Section 5 jurisdiction is not what legal form the entity takes, but rather whether the entity is "organized to carry on business for its own profit or that of its members."

Investor-Owned Utilities

Investor-owned utilities are clearly subject to the FTC's Section 5 jurisdiction as "corporations." The EIA defines investor-owned electric utilities as those that "have the fundamental objective of producing a profit for their investors" and distributing these profits as dividends or reinvesting them in the business.[260] These utilities satisfy the definition of "corporation" under the statute because they are companies organized to carry on business for the profit of their investors.[261]

Publicly Owned Utilities

It is unclear whether the FTC has Section 5 jurisdiction over publicly owned utilities. The agency probably lacks jurisdiction over these utilities if it characterizes them as "corporations," but it is possible that it may have jurisdiction over them if it characterizes them as "persons." Publicly owned utilities include "municipals, public utility districts and public power districts, State authorities, irrigation districts, and joint municipal action agencies."[262] The EIA describes these as "nonprofit government entities that are organized at either the local or State level," are exempt from state and federal income taxes, and "provide service to their communities and nearby consumers at cost."[263] In contrast to investor-owned utilities or cooperatively owned utilities, publicly owned utilities obtain capital by issuing debt rather than selling an ownership interest in the utility to investors or members.[264]

As "Corporations"

Publicly owned utilities probably do not fall within the FTC's Section 5 jurisdiction over "corporations" because they are not organized to carry on business for profit. Rather, governments form these utilities for the sole purpose of distributing electricity to consumers at

---

[258] Cal. Dental Ass'n v. FTC, 526 U.S. 756, 768 (1999) (emphasis added) (internal quotation marks omitted).

[259] *Id.* at 766-69.

[260] EIA ELECTRIC POWER OVERVIEW, *supra* note 254.

[261] Indeed, the FTC has asserted Section 5 jurisdiction over holding companies with investor-owned electric utility subsidiaries in the past. *See, e.g., DTE Energy Co.*, 131 F.T.C. 962 (May 15, 2001) (complaint); *CMS Energy Corp.*, 127 F.T.C. 827 (June 2, 1999) (complaint). *See also In re* DTE Energy Co., FTC File No. 001 0067 (May 15, 2001) (consent order); *In re* CMS Energy Corp., FTC File No. 991 0046 (June 2, 1999) (consent order).

[262] EIA ELECTRIC POWER OVERVIEW, *supra* note 254.

[263] *Id.*

[264] DAVID E. MCNABB, PUBLIC UTILITIES: MANAGEMENT CHALLENGES FOR THE 21ST CENTURY 165 (2005).

---

cost.[265] Significantly, when publicly owned utilities realize net income—that is, revenues they earn in excess of their expenses—they either (1) use it to finance their operations in lieu of issuing more debt,[266] or (2) transfer it to the general fund of the political subdivision that they serve.[267] These utilities typically lack investors or members to which they could distribute net income as dividends.[268] Thus, publicly owned utilities are probably not "organized to carry on business" for profit and are probably exempt from the FTC's Section 5 jurisdiction if characterized as "corporations."

As "Persons"

It is unclear whether a court would find that the FTC has Section 5 jurisdiction over publicly owned utilities as "persons," as a court could employ several different canons of statutory interpretation when deciding whether "persons" includes state or local government entities.[269] In the 1980s, the FTC attempted to assert Section 5 jurisdiction over two state-chartered municipal corporations—the cities of New Orleans and Minneapolis—as "persons," alleging that the cities engaged in unfair methods of competition by assisting taxicab companies in maintaining high prices and stifling competition.[270] The Commission later withdrew both complaints, and thus no court considered whether jurisdiction was proper. More recently, the Commission has asserted jurisdiction over state government agencies that regulate certain professions such as dentistry,[271] optometry,[272] and funeral services.[273]

There appears to be only one court case that engages in a full discussion and interpretation of the meaning of "persons" under Section 5. In *California State Board of Optometry v. FTC*, the D.C. Circuit Court of Appeals considered "whether a State acting in its sovereign capacity is a 'person' within the FTC's enforcement jurisdiction."[274] The FTC had issued a rule declaring "certain state laws restricting the practice of optometry to be unfair acts or practices."[275] Petitioners, which were state boards of optometry and professional associations, argued that the court should strike down the rule because it went beyond the FTC's statutory authority.[276] In vacating the rule, the court found nothing in the relevant provisions of the FTC Act "to indicate that Congress intended to authorize the FTC to reach the 'acts or practices' of States acting in their sovereign capacities."[277]

---

[265] EIA ELECTRIC POWER OVERVIEW, *supra* note 254.

[266] MCNABB, *supra* note 264, at 165.

[267] EIA ELECTRIC POWER OVERVIEW, *supra* note 254.

[268] MCNABB, *supra* note 264, at 165.

[269] In contrast to entities that are "corporations," the FTC does not have to show that entities qualifying as "persons" are organized for profit. *See* 15 U.S.C. §44.

[270] *In re* City of Minneapolis, 105 F.T.C. 304 (May 7, 1985) (order withdrawing complaint); *In re* City of New Orleans, 105 F.T.C. 1 (Jan. 3, 1985) (order withdrawing complaint).

[271] *In re* N.C. State Bd. of Dental Exam'rs, 151 F.T.C. 607 (Feb. 3, 2011) (state action opinion); *In re* South Carolina State Bd. of Dentistry, 138 F.T.C. 229 (Sept. 12, 2003) (complaint).

[272] *In re* Mass. Board of Registration in Optometry, 110 F.T.C. 549 (June 13, 1988) (decision).

[273] *In re* Va. Bd. of Funeral Dirs. & Embalmers, 138 F.T.C. 645 (Oct. 1, 2004) (complaint).

[274] 910 F.2d 976, 979 (D.C. Cir. 1990).

[275] *Id.* at 978.

[276] *Id.* at 978-79.

[277] *Id.* at 980, 982.

A court approaching the question of whether "persons" includes publicly owned utilities would start with the language of the statute. Courts traditionally give broad deference to an agency when the agency interprets the extent of its own jurisdiction unless the reach of its jurisdiction is clear from reading the statute "under ordinary principles of construction."[278] Attempting to discern the Commission's jurisdiction under Section 5 of the FTC Act is difficult, as the statute does not define the term "persons" for the purposes of that provision. Title 1, Section 1 of the United States Code (the Dictionary Act) provides: "In determining the meaning of any Act of Congress, *unless the context indicates otherwise* ... the words 'person' and 'whoever' include corporations, companies, associations, firms, partnerships, societies, and joint stock companies, as well as individuals."[279]

However, the context in which "persons" appears in Section 5 probably forecloses the use of the default definition of "person" in the Dictionary Act. In Section 5, Congress listed the terms "persons," "partnerships," and "corporations" separately, which indicates that it intended to give each term independent significance. The terms "corporations" and "partnerships" would not have independent meaning in Section 5 if the term "persons" in Section 5 included the entities listed in the Dictionary Act. Furthermore, the FTC Act requires that "corporations" be organized for their own profit or the profit of their members in order for the FTC to exercise jurisdiction over them—a requirement it does not impose on the other entities.[280] By reading the term "persons" to include the entities listed in the Dictionary Act, the FTC could evade this additional requirement simply by bringing its complaint against an entity as a "person" rather than a "corporation"—a result that Congress probably did not intend. Thus, a court that ended its analysis here could find that the meaning of "persons" remains ambiguous. The court could then choose to defer to the FTC's broad interpretation of its own jurisdiction under the Supreme Court's decision in *Chevron U.S.A., Inc. v. NRDC, Inc.*[281]

The *California Optometry* court, however, declined to defer to the FTC's interpretation of its own jurisdiction because it found that principles of federalism outweighed *Chevron* deference.[282] Quoting the Supreme Court's decision in *Will v. Michigan Department of State Police*,[283] the

---

[278] *See* Cal. Dental Ass'n v. FTC, 526 U.S. 756, 765-66 (1999) ("Respondent urges deference to this interpretation of the Commission's jurisdiction as reasonable. But we have no occasion to review the call for deference here, the interpretation urged in respondent's brief being clearly the better reading of the statute under ordinary principles of construction.") (internal citations omitted); *see also* Chevron U.S.A., Inc. v. NRDC, Inc., 467 U.S. 837, 842-43 (1984).

[279] 1 U.S.C. §1 (emphasis added).

[280] *See* 15 U.S.C. §44.

[281] *Chevron*, 467 U.S. at 842-43. In that case, the Court held that

> When a court reviews an agency's construction of the statute which it administers, it is confronted with two questions. First, always, is the question whether Congress has directly spoken to the precise question at issue. If the intent of Congress is clear, that is the end of the matter; for the court, as well as the agency, must give effect to the unambiguously expressed intent of Congress. If, however, the court determines Congress has not directly addressed the precise question at issue, the court does not simply impose its own construction on the statute, as would be necessary in the absence of an administrative interpretation. Rather, if the statute is silent or ambiguous with respect to the specific issue, the question for the court is whether the agency's answer is based on a permissible construction of the statute. *Id.*

[282] Todd H. Cohen, *Double Vision: The FTC, State Regulation, and Deciding What's Best for Consumers*, 59 GEO. WASH. L. REV. 1249, 1267 (1991) ("In sum, the *California State Board of Optometry* court relied on federalism principles to justify protecting state interests. The court extended the judicially-created *Parker* state action doctrine to cover FTC trade regulation rules and applied the clear statement doctrine to prevent the FTC from invalidating a state law as unfair without additional congressional action.").

[283] 491 U.S. 58 (1989).

California Optometry court stated that "in common usage, the term person does not include the sovereign, and statutes employing the word are ordinarily construed to exclude it."[284] In the *Will* case, the Court considered whether the term "person" as it appeared in 42 U.S.C. §1983 included a state.[285] The Court held that it did not, invoking the principles of federalism when it wrote that "[t]his approach is particularly applicable where it is claimed that Congress has subjected the States to liability to which they had not been subject before."[286] The Court found that the statute's language fell "far short of satisfying the ordinary rule of statutory construction that if Congress intends to alter the 'usual constitutional balance between the States and Federal Government,' it must make its intention to do so 'unmistakably clear in the language of the statute.'"[287]

The Court's decision in *Will*, as interpreted by the D.C. Circuit in *California Optometry*, suggests that Congress must clearly indicate in a particular statute when it wishes to subject states to a new form of liability, particularly when this would change the balance between state and federal authority by intruding on the actions a state takes in its sovereign capacity. There does not appear to be a clear indication that Congress intended the word "persons" in the FTC Act to subject publicly owned utilities to FTC enforcement actions.[288] Thus, if the FTC's enforcement of Section 5 against a publicly owned utility would alter the balance between the state and federal governments, a court might read "persons" to exclude these utilities. As the *California Optometry* court indicated, whether the balance is altered may depend on whether the operation of the utility amounts to the state acting in its sovereign capacity (balance altered) or merely engaging in a proprietary function (balance not altered).[289] The *California Optometry* court suggested that whether a state is acting in its sovereign capacity or engaging in a proprietary function may vary according to the antitrust laws' state action doctrine, a multi-pronged analysis that is beyond the scope of this report.[290] If a court found that the state was acting in its sovereign capacity when the state (or one of its subdivisions) operated an electric utility, the court could hold that the FTC does not have Section 5 jurisdiction because of the federalism principles and clear statement rule that guided the interpretation of the statute in *Will* and were adopted by the court in *California Optometry*.[291]

A third possible choice for a court would be to adopt the reasoning of the FTC and find that Congress clearly intended "persons" to include government entities, because under the other antitrust laws, the term "persons" includes state and local government entities, and the antitrust

---

[284] *California Optometry*, 910 F.2d 976, 980 (D.C. Cir. 1990) (internal quotation marks omitted).

[285] *Will*, 491 U.S. at 60.

[286] *Id.* at 64.

[287] *Id.* at 65 (citations omitted).

[288] Representative Covington, the sponsor of the act, explained during floor debate on the measure that Section 5 "embraces within the scope of that section every kind of person, natural or artificial, who may be engaged in interstate commerce."51 CONG. REC. 14,928 (1914). Despite this remark, courts have not taken such a broad view of the FTC's jurisdiction under the act. Even the Supreme Court has held that there are some limits on the entities covered by Section 5. *See* Cal. Dental Ass'n v. FTC, 526 U.S. 756, 766-67 (1999) (requiring, for jurisdiction, that a "proximate relation" must exist between the activities of a nonprofit and the benefit it provides to its members, and implying that the activities must confer "more than *de minimis* or merely presumed economic benefits" on the members).

[289] *See California Optometry*, 910 F.2d at 980-81 ("This rule of statutory construction serves to ensure that the States' sovereignty interests are adequately protected by the political process.").

[290] *Id.* at 980. For more information on the factors that courts consider when making this determination, see FED. TRADE COMM'N, REPORT OF THE STATE ACTION TASK FORCE (2003), *available at* http://www.ftc.gov/os/2003/09/ stateactionreport.pdf.

[291] *See* Cohen, *supra* note 282, at 1267.

laws, including the FTC Act,[292] should be read together.[293] The *California Optometry* court acknowledged this argument, writing that "several Supreme Court decisions hold that a State *is* a person for purposes of the antitrust laws."[294] The court ultimately rejected the argument, however, because it found that "when a State acts in a sovereign rather than a proprietary capacity, it is exempt from the antitrust laws even though those actions may restrain trade," and that this state action doctrine may "limit the reach of the FTC's enforcement jurisdiction."[295] Thus, if a court found that a state acted in its *proprietary* capacity when the state (or one of its subdivisions) operated a public utility, then the state action doctrine would not apply, and it would be possible for a court to find jurisdiction even under the *California Optometry* case. The FTC has advanced this reasoning, arguing that the state boards over which it asserts jurisdiction do not amount to the states acting in their sovereign capacities.[296] Whether the operation of a particular publicly owned utility consists of the state acting in its sovereign capacity or engaging in a proprietary function may vary according to the antitrust laws' state action doctrine, a multi-pronged analysis that is beyond the scope of this report.[297]

Thus, whether a court would find that the word "persons" in Section 5 includes certain government entities such as publicly owned utilities is unclear because it may depend on which, if any, of several principles of statutory construction the court adopts. A court could, among other options: (1) find that the meaning of "persons" in Section 5 is ambiguous, and thus defer to the FTC's broad interpretation of its own jurisdiction because of the *Chevron* doctrine; (2) find that the statute is ambiguous, but that principles of federalism outweigh the court's usual *Chevron* deference to the Commission's interpretation of its own jurisdiction—a determination that may require a court to find that the state is acting in its sovereign capacity when the state (or one of its subdivisions) operates an electric utility; or (3) find that Congress clearly intended "persons" to include government entities because Section 5 should be read together with the other antitrust laws, under which the term "person" includes state and local government entities—a determination that may require a court to find that the state is performing a proprietary function when the state (or one of its subdivisions) operates a utility.

Federally Owned Utilities

It is unclear whether the FTC could enforce Section 5 against a federally owned utility. Indeed, there does not appear to be any case in which the FTC has sought to enforce Section 5 against a federal agency.[298] The FTC probably lacks Section 5 jurisdiction over the nine federally owned

---

[292] Although this report focuses on the FTC's consumer law cases under Section 5 ("unfair or deceptive acts or practices"), and not its antitrust cases ("unfair methods of competition"), both types of prohibited activities share the same phrase for the purposes of determining the agency's jurisdiction: "persons, partnerships, or corporations." *See* 15 U.S.C. §45(a)(2).

[293] *See In re* Mass. Board of Registration in Optometry, 110 F.T.C. 549 (June 13, 1988) (decision) (citations omitted).

[294] *California Optometry*, 910 F.2d at 980 (citations omitted).

[295] *Id.* at 980 (citation omitted).

[296] *See, e.g., In re* N.C. State Bd. of Dental Exam'rs, 151 F.T.C. 607 (Feb. 3, 2011) (state action opinion); *In re* Mass. Board of Registration in Optometry, 110 F.T.C. 549 (June 13, 1988) (decision).

[297] For more information on the factors that courts consider when making this determination, see FED. TRADE COMM'N, REPORT OF THE STATE ACTION TASK FORCE (2003), *available at* http://www.ftc.gov/os/2003/09/stateactionreport.pdf.

[298] This report does not consider whether any constitutional implications would result if the FTC, an independent executive branch agency, brought an enforcement proceeding against another executive branch agency. *See generally* Michael Eric Herz, *When Can the Federal Government Sue Itself?*, 32 WM. & MARY L. REV. 893 (1991).

---

utilities operating in the United States[299] if it characterizes them as "corporations." Like publicly owned utilities, federally owned utilities are not organized for profit. As the EIA notes, "federal power is not sold for profit, but to recover the costs of operations and repay the Treasury for funds borrowed to construct generation and transmission facilities."[300] If the Commission characterizes these utilities as "persons," it is unclear whether a court would find that this term includes government entities.[301]

As a practical matter, FTC enforcement of Section 5 against federally owned utilities is probably unnecessary in the context of smart meter data because of other federal laws, such as the Privacy Act,[302] that would likely protect this data when it is stored in records systems maintained by federal agencies, including federally owned utilities.[303]

Cooperatively Owned Utilities

For-profit electric cooperatives would clearly fall within the Commission's Section 5 jurisdiction over "corporations" operated for their own profit or that of their members.[304] Indeed, the FTC has maintained jurisdiction over for-profit cooperatives as "corporations" in the past, including a rural healthcare cooperative[305] and a wine maker.[306] However, it appears that most electric cooperatives—and particularly the cooperatives that will receive funds under the Department of Energy's Smart Grid Investment Grant program—are nonprofits.[307]

It is possible that the FTC would have Section 5 jurisdiction over these nonprofit electric cooperatives as "corporations" organized for profit. These distribution utilities are owned by the "consumers they serve," and those that are tax-exempt must "provide electric service to their members at cost, as that term is defined by the Internal Revenue Service."[308] However, when the activities of a cooperative result in revenues that exceed the cooperative's costs, these "net margins ... are considered a contribution of equity by the members that are required to be returned to the members consistent with the organization's bylaws and lender limitations imposed as a condition of loans."[309] Thus, in contrast to publicly owned utilities, which typically transfer any net income to the general fund of the government that they serve, electric cooperatives return net margins to their members as equity, and when that equity is retired by the board of directors, members receive cash payments.[310] Although it does not appear that a court has considered

---

[299] EIA ELECTRIC POWER OVERVIEW, *supra* note 254. Among these utilities are the Tennessee Valley Authority, the four power marketing administrations in the Department of Energy, and the Army Corps of Engineers. *Id.*

[300] *Id.*

[301] *See supra* notes 269-97 and accompanying text.

[302] 5 U.S.C. §552a.

[303] *See* "The Federal Privacy Act of 1974," *infra* p. 45.

[304] 15 U.S.C. §44.

[305] *In re* Minn. Rural Health Coop., FTC File No. 051 0199 (Dec. 28, 2010) (decision and order).

[306] *In re* Heublein, Inc., 96 F.T.C. 385 (Oct. 7, 1980) (final order).

[307] *See* DEP'T OF ENERGY, CASE STUDY – NATIONAL RURAL ELECTRIC COOPERATIVE ASSOCIATION SMART GRID INVESTMENT GRANT 1, *available at* http://energy.gov/sites/prod/files/oeprod/DocumentsandMedia/ NRECA_case_study.pdf.

[308] EIA ELECTRIC POWER OVERVIEW, *supra* note 254.

[309] *Id.* "Net margins" is the term given to "revenues in excess of the cost of providing service." *Id.*

[310] *See, e.g.*, Cent. Rural Electric Coop., Patronage Capital, http://www.crec.coop/CRECAdvantage/PatronageCapital/ tabid/711/Default.aspx ("Allocated patronage capital appears as an entry on the permanent financial records of the (continued...)

---

whether the FTC has Section 5 jurisdiction over a nonprofit electric cooperative that returns its net margins to its consumer-members in addition to providing them with electricity service, the Supreme Court, as well as lower federal courts, have issued guidance on factors that a court may consider in answering this question.

Applicable Law

Under Section 5, the FTC Act requires that a "corporation" be "organized to carry on business for its own profit *or that of its members*."[311] In *California Dental Ass'n v. FTC*, the Court considered whether the FTC could enforce Section 5 against a "voluntary nonprofit association of local dental societies" that was exempt from paying federal income tax and furnished its members with "advantageous insurance and preferential financing arrangements" in addition to lobbying, litigating, and advertising on their behalf.[312] The Court found that the FTC had jurisdiction over the California Dental Association as a "corporation," stating that

> the FTC Act is at pains to include not only an entity "organized to carry on business for its own profit," but also one that carries on business for the profit "of its members." While such a supportive organization may be devoted to helping its members in ways beyond immediate enhancement of profit, no one here has claimed that such an entity must devote itself single-mindedly to the profit of others. It could, indeed, hardly be supposed that Congress intended such a restricted notion of covered supporting organizations, with the opportunity this would bring with it for avoiding jurisdiction where the purposes of the FTC Act would obviously call for asserting it.[313]

The Court declined to specify the percentage of a nonprofit entity's activities that must be "aimed at its members' pecuniary benefit" to subject it to FTC jurisdiction.[314] However, the Court wrote that a "proximate relation" must exist between the activities of the entity and the profits of its members, and implied that the activities must confer "more than *de minimis* or merely presumed economic benefits" on the members.[315] The Court's justification for this result was that "nonprofit entities organized on behalf of for-profit members have the same capacity and derivatively, at

---

(...continued)

cooperative and reflect [sic] your equity or ownership in CREC. When patronage capital is retired, a check or bill credit is issued to you and your equity in the cooperative is reduced. ... When considering a retirement, the board analyzes the financial health of the cooperative and will not authorize a retirement that will adversely affect the financial integrity of the cooperative."); Fall River Rural Electric Coop., Patronage Capital, http://www.frrec.com/myAccount/ patronageCapital.aspx ("The Cooperative's Board of Directors retires patronage capital when finances allow, often on an annual basis. The oldest patronage capital is retired first. Fall River currently retires patronage capital on a rotation of approximately 20 years."); Kauai Island Util. Coop., Member Patronage Capital Information, http://www kiuc.coop/ member_patcap-qa htm ("A portion of Patronage Capital may be periodically paid to the members upon approval of the Board of Directors and our lenders."); Sulphur Springs Valley Electric Coop., Inc., Patronage Capital Credits, http://www.ssvec.org/?page_id=583 ("Capital credits represent your share of the Cooperative's margins – margins are the operating revenue remaining after operating expenses. The amount assigned in your name depends on your energy purchases. To calculate this, we divide your annual energy purchase by the Cooperative's operating income for the year. The more electricity you buy, the more capital credits you earn.").

[311] 15 U.S.C. §44 (emphasis added).

[312] 526 U.S. 756, 759-60, 767 (1999).

[313] *Id.* at 766 (internal citations omitted).

[314] *Id.*

[315] *Id.* at 766-67.

least, the same incentives as for-profit organizations to engage in unfair methods of competition or unfair and deceptive acts."[316]

It is clear that the FTC may still have Section 5 jurisdiction even when the benefits that a nonprofit provides to its members are secondary to its charitable functions. In *American Medical Ass'n v. FTC*, the Second Circuit considered whether the FTC could enforce Section 5 against three medical professional associations, including the American Medical Association (AMA), a nonprofit corporation composed of "physicians, osteopaths, and medical students."[317] The court, acknowledging that the associations served "both the business and non-business interests of their member physicians," found jurisdiction because the "business aspects" of their activities, including lobbying for members and offering business advice to them, subjected them to the FTC's jurisdiction despite the fact that the business aspects "were considered secondary to the charitable and social aspects of their work."[318]

When determining whether jurisdiction exists, a court may consider other factors in addition to the benefits that the nonprofit provides to its members. In *Community Blood Bank v. FTC*, the Eighth Circuit considered whether a "corporation" included all nonprofit corporations.[319] The appeals court held that the FTC lacked Section 5 jurisdiction over nonprofit blood banks because the banks' activities did not result in "profit" in the sense of "gain from business or investment over and above expenditures."[320] The blood banks, the court observed, lacked shares of capital, capital stock, or certificates, and were "organized for and actually engaged in business for only charitable purposes."[321] One bank's articles of incorporation touted the entity's charitable purposes, and all of the banks were exempt from paying federal income taxes.[322] Upon dissolution, the corporations would transfer their assets to other charitable or nonprofit organizations.[323] In addition, none of the funds collected by the blood banks had "ever been distributed or inured to the benefit of any of their members, directors or officers."[324] The court found that these factors made the blood banks "charitable organizations" both "in law and in fact," exempting them from the FTC's Section 5 jurisdiction.[325]

Analysis

The case law suggests several factors that a court may weigh when determining whether a private, nonprofit entity composed of members, such as an electric cooperative, is subject to the FTC's Section 5 jurisdiction as a "corporation."[326] The most significant factor is whether the nonprofit

---

[316] *Id.* at 768.

[317] 638 F.2d 443, 446 (1980).

[318] *Id.* at 448. The court noted in passing that the AMA's articles of incorporation stated that one purpose of the organization was to "safeguard the material interests of the medical profession." *Id.*

[319] 405 F.2d 1011, 1015 (8th Cir. 1969).

[320] *See id.* at 1017. The court also remarked that at least one case had established that "even though a corporation's income exceeds its disbursements its nonprofit character is not necessarily destroyed." *Id.*

[321] *Id.* at 1020, 1022.

[322] *Id.* at 1020.

[323] *Id.*

[324] *Id.*

[325] *Id.* at 1019.

[326] This analysis assumes that a court would extend the holdings of the applicable case law, which covered entities organized as nonprofit corporations and professional associations, to include entities organized as nonprofit electric (continued...)

---

provides an economic benefit to its members that is more than *de minimis* and that is proximately related to the nonprofit's activities. This benefit need not be the sole—or even primary—function of the nonprofit. Additional factors that the case law suggests weigh in favor of a finding of jurisdiction include that the nonprofit: (1) has gain from its business or investments that exceeds its expenditures; (2) has shares of capital or capital stock or certificates; (3) is not organized solely for charitable purposes or does not engage only in charitable work; (4) has articles of incorporation that list profit-seeking objectives; (5) is subject to federal income tax liability; (6) would distribute its assets to profit-seeking entities upon dissolution; and (7) distributes any of the funds it collects to its members, directors, or officers.

It is possible that the FTC has Section 5 jurisdiction over nonprofit electric cooperatives, although the outcome in any particular case may depend on the characteristics of the individual utility. A court could find that the typical nonprofit electric cooperative provides "economic benefit" to its members in at least two ways: (a) by providing electricity service to members;[327] and (b) by returning net margins to members in the form of patronage capital, which is an ownership interest in the cooperative that is later converted to cash payments to members when that capital is retired.[328] With regard to (a), it is likely that a court would find that electricity service is an "economic benefit" as defined in the case law. In *California Dental Ass'n*, the nonprofit professional association provided "advantageous insurance and preferential financing arrangements," as well as lobbying, litigation, and advertising services to its members.[329] In *American Medical Ass'n*, the nonprofit lobbied on behalf of its members and offered business advice to members.[330] These benefits, it is assumed, enabled the members to more easily conduct business profitably. Electricity service allows people to conduct activities at all times of the day, and thus provides a similar and clearly significant economic benefit to those who use it, whether for business or recreational purposes. As the primary objective of an electric cooperative is to provide electricity service to members, the necessary proximate relation between the activities of the nonprofit and the benefit to its members clearly exists.

Despite its pecuniary nature, there are a few problems with considering benefit (b), patronage capital, to be an "economic benefit" as defined by the Court. First, it is not clear that patronage capital actually is a benefit. A court could view patronage capital as a no-interest *loan* from the consumer-member to the utility,[331] or, because it is typically allocated to member accounts in a manner proportional to members' spending on electricity, simply a *refund* of money collected from the members that reflects the actual cost of providing service in a particular year.[332] If

---

(...continued)
cooperatives.

[327] Many cooperatives provide other services to their communities that could constitute "economic benefits." The National Rural Electric Cooperative Association notes that, "In addition to electric service, many electric co-ops are involved in community development and revitalization projects" that include "small business development and jobs creation, improvement of water and sewer systems, and assistance in delivery of health care and educational services." Nat'l Rural Electric Coop. Ass'n, Member Directory, http://www.nreca.coop/members/MemberDirectory/Pages/default.aspx.

[328] *See* sources cited *supra* note 310.

[329] Cal. Dental Ass'n v. FTC, 526 U.S. 756, 759-60, 767 (1999).

[330] Am. Med. Ass'n v. FTC, 638 F.2d 443, 448 (1980).

[331] *See, e.g.*, Cent. Rural Electric Coop., Patronage Capital, http://www.crec.coop/CRECAdvantage/PatronageCapital/tabid/711/Default.aspx ("These margins represent an interest-free loan of operating capital by the membership to the cooperative.").

[332] *See, e.g.*, Kauai Island Util. Coop., Member Patronage Capital Information, http://www.kiuc.coop/member_patcap-
(continued...)

---

adopted by a court, neither of these characterizations would appear to be consistent with the "profit" that the statute describes[333] or the "economic benefit" that the Supreme Court requires for a nonprofit to be a "corporation."

Second, even if a court found patronage capital to be an economic benefit, it is not clear that it is more than *de minimis.* Patronage capital must be "retired" before members receive cash payments for it.[334] Retirements are made at the discretion of the cooperative's board of directors because the capital is needed to finance the cooperative's ongoing expenses, and thus retirement of a class of capital typically occurs after a long rotation period, such as 20 years.[335] Although the Supreme Court did not hold that an "economic benefit" must produce *immediate* advantage to the members of a nonprofit, a court could potentially view the decades-long delay in cash payments as significantly decreasing the degree of economic benefit that the capital provides. In addition, patronage capital would probably be considered *de minimis* if the cooperative's net margins were small, as this would mean that little capital would be issued to members. It is thus difficult to discern whether a court would find that an economic benefit accrues to members as a result of their receipt of patronage capital, which nevertheless probably bears the requisite "proximate relation" to the activities of the cooperative that produce any net margins distributed as capital.

With regard to the additional factors, those favoring jurisdiction include (2) cooperatives typically have shares of capital stock, including patronage capital;[336] (3) cooperatives do not operate solely for the benefit of the people outside of the organization like the nonprofits in *Community Blood Bank* did because cooperatives provide electricity service and patronage capital to their members;[337] and (7) an electric cooperative typically returns any net margins to members in the form of patronage capital, an ownership interest refunded to consumer-members as cash when the capital is retired.[338] Factors that cannot be evaluated because they are specific to each individual cooperative include (1) whether the revenues of the cooperative exceed its expenditures; (4) the particular objectives listed in a cooperative's articles of incorporation or other foundational document; (5) whether a nonprofit electric cooperative is exempt from federal income tax liability, which depends on whether it meets the requirements under Section 501(c)(12) of the Internal Revenue Code;[339] and (6) whether a cooperative would distribute its assets to profit-seeking entities upon dissolution—a factor that also may depend on state laws.[340]

It is likely that a court would find that nonprofit electric cooperatives impart economic benefits to their members by distributing electricity to them or, possibly, by issuing patronage capital to them. However, because many of the other factors that courts consider may differ for each

---

(...continued)

qa htm (characterizing the retirement of patronage capital as a "refund").

[333] 15 U.S.C. §44.

[334] *See* sources cited *supra* note 310.

[335] *See id.*

[336] *See* Nat'l Rural Electric Coop. Ass'n, Seven Cooperative Principles, http://www.nreca.coop/members/ SevenCoopPrinciples/Pages/default.aspx (describing "Members' Economic Participation").

[337] Whether electricity service and patronage capital, which are clearly benefits, constitute "economic benefits" within the meaning of the Supreme Court's holding in *California Dental Ass'n* is a separate question.

[338] *See* sources cited *supra* note 310.

[339] I.R.C. §501(c)(12).

[340] *See* Cmty. Blood Bank v. FTC, 405 F.2d 1011, 1020 (8th Cir. 1969).

particular cooperative, it is not possible to draw any general conclusions about whether the FTC would have Section 5 jurisdiction over these entities as "corporations."

Enforcement of Data Privacy and Security

If the FTC has Section 5 jurisdiction over a particular electric utility, it may bring an enforcement action against the utility if its privacy or security practices with regard to consumer smart meter data constitute "unfair or deceptive acts or practices in or affecting commerce."[341] The FTC Act defines an "unfair" act or practice as one that "causes or is likely to cause substantial injury to consumers which is not reasonably avoidable by consumers themselves and not outweighed by countervailing benefits to consumers or to competition."[342] According to the FTC, an act or practice is "deceptive" if it is a material "representation, omission or practice" that is likely to mislead a consumer acting reasonably in the circumstances.[343] The history of the Commission's enforcement of consumer data privacy and security practices shows that the agency has brought complaints against entities that (1) engage in "deceptive" acts or practices by failing to comply with their stated privacy policies; or (2) employ "unfair" practices by failing to adequately secure consumer data from unauthorized parties.[344] Often, conduct constituting a violation could fall under either category, as a failure to protect consumer data may be an unfair practice because of the unavoidable injury it causes, as well as a deceptive practice because it renders an entity's privacy policy materially misleading.

"Deceptive" Privacy Statements

A utility that fails to comply with its own privacy policy may engage in a "deceptive" act or practice under Section 5 of the FTC Act. In *Facebook, Inc.*, the FTC alleged, among other things, that the social networking site violated promises contained in its privacy policy when it made users' personal information accessible to third parties without users' consent.[345] Facebook had claimed that users could limit third-party access to their personal information on the site. Despite this promise, applications run by users' Facebook friends were able to access the users' personal information. The Commission also charged that Facebook altered its privacy practices without users' consent, causing personal information that had been restricted by users to be available to third parties. This change, which allegedly "caused harm to users, including, but not limited to, threats to their health and safety, and unauthorized revelation of their affiliations" constituted both a "deceptive" and an "unfair" practice in the view of the Commission.[346] Finally, the Commission alleged that Facebook had represented to users that it would not share their personal information with advertisers but had done so anyway.

---

[341] 15 U.S.C. §45(a)(1). For more details on FTC enforcement of consumer data privacy and security under Section 5, see CRS Report RL34120, *Federal Information Security and Data Breach Notification Laws*, by Gina Stevens.

[342] 15 U.S.C. §45(n).

[343] *In re* Cliffdale Assocs., Inc., 103 F.T.C. 110, 174 (1984) (policy statement at end of opinion).

[344] *See Consumer Privacy: Hearing Before the S. Comm. on Commerce, Sci., and Transp.*, 11th Cong. (2010) (statement of Jon D. Leibowitz, Chairman, Fed. Trade Comm'n) (describing the FTC's enforcement activity in the areas of consumer data privacy and security), *available at* http://www.ftc.gov/os/testimony/100727consumerprivacy.pdf. The FTC recently released a preliminary report on the consumer privacy implications of new technologies. FED. TRADE COMM'N, PROTECTING CONSUMER PRIVACY IN AN ERA OF RAPID CHANGE: A PROPOSED FRAMEWORK FOR BUSINESSES AND POLICYMAKERS (2010), *available at* http://www.ftc.gov/os/2010/12/101201privacyreport.pdf.

[345] FTC File No. 092 3184 (Nov. 29, 2011) (complaint).

[346] *Id.*

In *Twitter, Inc.*, the FTC alleged that the social networking site engaged in "deceptive" acts when it violated claims made in its privacy policy about the security of consumer data by failing to "use reasonable and appropriate security measures to prevent unauthorized access to nonpublic user information."[347] The Commission found that Twitter had permitted its administrators to access the site with easy-to-guess passwords and failed to limit the extent of administrators' access according to the requirements of their jobs. In a consent order, the company agreed not to misrepresent its privacy controls and to implement a comprehensive information security program that would be assessed by an independent third party.[348]

As smart meter data becomes valuable to third parties,[349] utilities may be tempted to sell or share this information with others to increase revenues and provide new services to their customers. If prohibited by the terms of the utility's privacy policy, it may be a "deceptive" act or practice for the utility to share a consumer's personal information with third parties without a consumer's consent.[350] The FTC could also find deception when a utility represents that its privacy controls are capable of protecting smart meter data when, in fact, they are not.

"Unfair" Failure to Secure Consumer Data

Failure to Protect Against Common Technology Threats or Unauthorized Access

The FTC may consider it an "unfair" practice when an electric utility fails to safeguard smart meter data from well-known technology threats as the data travels across the utility's communications networks. For example, in *DSW Inc.*, the FTC brought enforcement proceedings against the respondent, the owner of several shoe stores.[351] The agency alleged that the respondent failed to protect customers' credit card and check information as it was transmitted to the issuing bank for authorization. The information collected at the register traveled wirelessly to the store's computer network, and from there to the bank or check processor, which communicated its response back to the store through the same channels. The agency charged that

> [a]mong other things, respondent (1) created unnecessary risks to the information by storing it in multiple files when it no longer had a business need to keep the information; (2) did not use readily available security measures to limit access to its computer networks through wireless access points on the networks; (3) stored the information in unencrypted files that could be accessed easily by using a commonly known user ID and password; (4) did not limit sufficiently the ability of computers on one in-store network to connect to computers on other in-store and corporate networks; and (5) failed to employ sufficient measures to detect unauthorized access. As a result, a hacker could use the wireless access points on one in-store computer network to connect to, and access personal information, on the other in-store and corporate networks.[352]

---

[347] FTC File No. 092 3093 (Mar. 2, 2011) (complaint).

[348] FTC File No. 092 3093 (Mar. 2, 2011) (decision and order)

[349] NIST PRIVACY REPORT, *supra* note 11, at 14, 35-36.

[350] As suggested below, it may also be an "unfair" practice, regardless of whether the utility has a privacy policy.

[351] FTC File No. 052 3096 (Mar. 7, 2006) (complaint).

[352] *Id.*

Similarly, in *Cardsystems Solutions, Inc.*, the Commission brought a complaint against a credit and debit card authorization processor.[353] The FTC alleged that the respondent failed to protect its systems by neglecting to guard its network against "commonly known or reasonably foreseeable attacks" that could be avoided using low-cost methods.[354] As part of settlement agreements in *DSW* and *Cardsystems*, the respondents had to create "a comprehensive information security program" to protect consumer information that would be assessed periodically by an independent third party.[355]

Smart meters also transmit personal consumer information, often wirelessly, across several different communications networks located in various physical places.[356] Thus, it is possible that the FTC would view a utility's failure to protect smart meter data against common technology threats as an "unfair" practice if the utility could have avoided the threats by using low-cost methods such as encrypting the data; storing it in fewer places and for no longer than needed; implementing basic wireless network security; and taking other reasonable measures suggested by the agency in *DSW Inc.*

Failure to Dispose of Data Safely

A utility's failure to dispose of smart meter data safely may also constitute an "unfair" practice under Section 5. For example, in *Rite Aid Corp.*, the respondent, the owner of retail pharmacy stores, purportedly failed to safely dispose of personal information in its possession when it neglected to: "(1) implement policies and procedures to dispose securely of such information," including rendering "the information unreadable in the course of disposal; (2) adequately train employees to dispose securely of such information; (3) use reasonable measures to assess compliance with its established policies and procedures for the disposal of such information; and (4) employ a reasonable process for discovering and remedying risks to such information."[357] The information was later found in various publicly accessible garbage dumpsters in readable form. This suggests that utilities holding smart meter data and other personal information, whether on electronic or physical media, must ensure that the methods used to destroy this data render it unreadable to third parties.

Penalties

There is no private right of action in the FTC Act. If the Commission has "reason to believe" that a violation has occurred, it may, after notice to the respondent and an opportunity for a hearing, issue an order directing the respondent to cease and desist from acts or practices that the agency finds violate the act.[358] If the respondent disobeys an order that has become final, the U.S. Attorney General may bring an action in district court seeking the imposition of civil monetary

---

[353] FTC File No. 052 3148 (Sept. 5, 2006) (complaint).

[354] *Id.*

[355] *See, e.g., In re* Cardsystems Solutions, Inc., FTC File No. 052 3148 (Sept. 5, 2006) (decision and order).

[356] NIST PRIVACY REPORT, *supra* note 11, at 23.

[357] FTC File No. 072 3121 (Nov. 12, 2010) (complaint).

[358] 15 U.S.C. §45(b). The Commission may seek a preliminary injunction in district court if it "has reason to believe" that an entity subject to the Commission's jurisdiction "is violating, or is about to violate, any provision of law enforced" by the FTC, and such an injunction would be in the public interest. 15 U.S.C. §53(b). In "proper cases the Commission may seek, and after proper proof, the court may issue, a permanent injunction." *Id.*

penalties of up to $16,000 per violation ($16,000 per day for continuing violations), as well as further injunctive and equitable relief that the court deems appropriate.[359]

After a party becomes subject to a final cease and desist order under the act, the Commission may seek redress for consumers by bringing suit in state or federal court against the party if the Commission "satisfies the court that the act or practice to which the cease and desist order relates is one which a reasonable man would have known under the circumstances was dishonest or fraudulent."[360] "Such relief may include, but shall not be limited to, rescission or reformation of contracts, the refund of money or return of property, the payment of damages," and public notification of the violation, "except nothing in [15 U.S.C. §57b(b)] is intended to authorize the imposition of any exemplary or punitive damages."[361] Once the Commission has issued a final cease and desist order (not a consent order) finding an act or practice to be deceptive, then it may bring suit in district court to obtain a civil penalty against an entity that engages in that act or practice: (1) after the order became final ("whether or not such person, partnership, or corporation was subject to such cease and desist order"); and (2) "with actual knowledge that such act or practice is unfair or deceptive and is unlawful" under Section 5 of the FTC Act.[362]

The Federal Privacy Act of 1974 (FPA)

Smart meter electricity usage data pertaining to U.S. citizens or permanent residents that is retrievable by personal identifier from a system of records maintained by any federal "agency," including federally owned utilities, is subject to the protections contained in the Privacy Act[363] when it is maintained, collected, used, or disseminated by the agency.

Federally Owned Utilities as "Agencies"

All nine of the federally owned utilities are federal agencies covered by the Privacy Act. For the purposes of the act, the term "agency" includes, but is not limited to, "any executive department, military department, Government corporation, Government controlled corporation, or other establishment in the executive branch of the Government (including the Executive Office of the President), or any independent regulatory agency."[364] According to EIA, utilities that are part of an executive department include the four power marketing administrations in the Department of Energy (Bonneville, Southeastern, Southwestern, and Western), the International Boundary and Water Commission in the Department of State, and the Bureau of Indian Affairs and the Bureau

---

[359] 15 U.S.C. §45(l). The size of the civil monetary penalty was last adjusted for inflation in 2009. 16 C.F.R. §1.98.

[360] 15 U.S.C. §57b(a)(2).

[361] 15 U.S.C. §57b(b).

[362] 15 U.S.C. §45(m)(1)(B).

[363] 5 U.S.C. §552a. The federally owned utilities primarily sell electricity to nonprofit electric utilities on the wholesale markets rather than distribute electricity directly to consumers. EIA ELECTRIC POWER OVERVIEW, *supra* note 254. As these utilities provide only about 1% of total sales of electricity to end user consumers, *id.*, they may be unlikely to acquire consumer smart meter data, which is typically transmitted to distribution utilities. However, as the smart grid becomes more interconnected, more utilities at different points in the smart grid may come into possession of this data. NIST PRIVACY REPORT, *supra* note 11, at 23.

[364] *See* 5 U.S.C. §552(f)(1). The act also covers data in a "system of records" operated by a government contractor on behalf of a federal agency. *See* 5 U.S.C. §552a(m).

---

of Reclamation in the Department of the Interior.[365] The U.S. Army Corps of Engineers resides in the Department of Defense, which is an executive department.[366] The Tennessee Valley Authority is a government-owned corporation.[367]

Smart Meter Data as a Protected "Record"

The Privacy Act protects the type of electricity usage data gathered by smart meters, provided that the data pertains to U.S. citizens or permanent residents, is personally identifiable, and is retrievable by the individual's name or another personal identifier. The Privacy Act "governs the collection, use, and dissemination of a 'record' about an 'individual' maintained by federal agencies in a 'system of records.'"[368] Under the statute, a "record" is "any item, collection, or grouping of information about an individual that is maintained by an agency ... that contains his name, or the identifying number, symbol, or other identifying particular assigned to the individual, such as a finger or voice print or a photograph."[369]

An "individual" is defined as "a citizen of the United States or an alien lawfully admitted for permanent residence."[370] A "system of records" is "a group of any records under the control of any agency from which information is retrieved by the name of the individual" or other personal identifier "assigned to the individual."[371]

Smart meter data held by an agency certainly fits within the broad definition of a "record" because it is a grouping of information about an individual, namely, data on that individual's electricity usage. The data is typically stored along with a consumer's account information, which usually includes a consumer's name, social security number, or other "identifying particular."[372] Thus, smart meter data would constitute a protected "record" under the Privacy Act, assuming that it pertains to a citizen of the United States or lawful permanent resident and is retrievable by a personal identifier such as a consumer's name or account number.

Requirements

For information on the general safeguards that the Privacy Act provides for data that is maintained by a federal agency and meets the other requirements for a covered record under the act, see CRS Report RL34120, *Federal Information Security and Data Breach Notification Laws*, by Gina Stevens.

---

[365] EIA ELECTRIC POWER OVERVIEW, *supra* note 254.

[366] DEP'T OF THE ARMY CORPS OF ENG'RS, CIVIL WORKS STRATEGIC PLAN 1 (2004), *available at* http://www.corpsresults.us/pdfs/cw_strat.pdf. It is also a "Major Command within the Army." *Id.*

[367] Tenn. Valley Auth., About TVA, http://www.tva.com/abouttva/index htm.

[368] *See* CRS Report RL34120, *Federal Information Security and Data Breach Notification Laws*, by Gina Stevens (citations omitted).

[369] 5 U.S.C. §552(a)(4).

[370] 5 U.S.C. §552a(a)(2).

[371] 5 U.S.C. §552a(a)(5).

[372] NIST PRIVACY REPORT, *supra* note 11, at 26-27.

Author Contact Information

Brandon J. Murrill
Legislative Attorney
bmurrill@crs.loc.gov, 7-8440

Edward C. Liu
Legislative Attorney
eliu@crs.loc.gov, 7-9166

Richard M. Thompson II
Legislative Attorney
rthompson@crs.loc.gov, 7-8449

www.ingramcontent.com/pod-product-compliance
Lightning Source LLC
Chambersburg PA
CBHW081356170526
45166CB00010B/3108